Plant Resources of South-East Asia No 1

Pulses

Plant Resources of South-East Asia is a multivolume handbook that aims to summarize knowledge about useful plants for workers in education, research, extension and industry. The following institutions are responsible for the coordination of the Prosea Programme and the handbook:

- Forest Research Institute of Malaysia (FRIM), Karung Berkunci 201, Jalan FRI Kepong, 52109 Kuala Lumpur, Malaysia
- Indonesian Institute of Sciences (LIPI), Widya Graha, Jalan Gatot Subroto 10, Jakarta 12710, Indonesia
- Institute of Ecology and Biological Resources, National Centre for Scientific Research (IEBR-NCSR), Nghia Do, Tu Liem, Hanoi, Vietnam
- Papua New Guinea University of Technology (UNITECH), Private Mail Bag, Lae, Papua New Guinea
- Philippine Council for Agriculture, Forestry and Natural Resources Research & Development (PCARRD), Los Banos, Laguna, the Philippines
- Thailand Institute of Scientific and Technological Research (TISTR), 196 Phahonyothin Road, Bang Khen, Bangkok 10900, Thailand
- Wageningen Agricultural University (WAU), Costerweg 50, 6701 BH Wageningen, the Netherlands

Plant Resources of South-East Asia

No 1

Pulses

L.J.G. van der Maesen and
Sadikin Somaatmadja (Editors)

PROSEA

Bogor Indonesia 1992

Professor Dr L.J.G. van der Maesen is a tropical agronomist-botanist who graduated from Wageningen Agricultural University in 1968. His thesis, defended in 1972, was a monograph on *Cicer*. He worked in Iraq for FAO and was a principal germplasm botanist with ICRISAT, India, from 1974 to 1984. In 1984, he joined the staff of the Department of Plant Taxonomy as professor and head of department. His publications cover various aspects of genetic resources and their collection, conservation and use, in particular of pulses. He has carried out several collection trips in Asia and Africa.
Dr Sadikin Somaatmadja is a plant breeder, who gained his M.Sc. in agronomy from Mississippi State University, United States, in 1963, and his Ph.D. in agriculture from the University of Tokyo, Japan, in 1988. He has been working at the Research Institute for Food Crops since 1951 and was Director of the Sukamandi branch of the Central Research Institute for Agriculture (CRIA) from 1971–1979. He has developed several soya bean cultivars. He has published mainly on soya bean and other major legumes in Indonesia. Since 1979, he has been Coordinator for Legume Crops at the Central Research Institute for Food Crops (CRIFC), Bogor, and since 1987 Regional Coordinator of the FAO/UNDP Project RAS/82/002 Technical Cooperation Among Developing Countries (TCDC) for the Research and Development of Food Legumes and Coarse Grains in the Tropics and Sub-Tropics of Asia.

ISBN 979-8316-00-2 (series)
ISBN 979-8316-01-0

Design: Frits Stoepman GVN.

To
W. M. Otto and
Soekiman Atmosoedaryo

Contents

Editors and contributors

Editorial staff of this volume

- Editors: L.J.G. van der Maesen and Sadikin Somaatmadja
- General editors: E. Westphal and P.C.M. Jansen
- Associate editors: R.K. Arora and J.S. Siemonsma
- Illustrators: P. Verheij-Hayes (drawings) and R. Boekelman (map)
- Publishing consultant: R.J.P. Aalpol
- In-house editor: J.C. Rigg

Contributors

- R.K. Arora, National Bureau of Plant Genetic Resources, New Delhi, India
- Arwooth Na Lampang, Department of Agriculture, Bangkhen, Bangkok, Thailand
- J.P. Baudoin, Faculté des Sciences Agronomiques de l'Etat, Phytotechnie des Régions Chaudes, Gembloux, Belgium
- D.R. Davies, John Innes Institute and AFRC Institute of Plant Science Research, Norwich, United Kingdom
- P.C.M. Jansen, Prosea Project, Department of Plant Taxonomy, Wageningen Agricultural University, the Netherlands
- R.S. Kulkarni, University of Agricultural Sciences, Bangalore, India
- A.R. Linnemann, Department of Tropical Crop Science, Wageningen Agricultural University, the Netherlands
- L.J.G. van der Maesen, Department of Plant Taxonomy, Wageningen Agricultural University, the Netherlands
- Shri S. Mauria, National Bureau of Plant Genetic Resources, New Delhi, India
- C.C.C.M. van Oers, Department of Tropical Crop Science, Wageningen Agricultural University, the Netherlands
- R.K. Pandey, International Rice Research Institute, Rice Farming Systems, Manila, Philippines
- A. Patanothai, Department of Plant Science, Faculty of Agriculture, Khon Kaen University, Thailand
- Sadikin Somaatmadja, Agency for Agricultural Research and Development, Central Research Institute for Food Crops, Bogor, Indonesia
- S. Shanmugasundaram, Asian Vegetable Research and Development Center, Crop Improvement Program, Shanhua, Tainan, Taiwan, Republic of China
- G. Shivashankar, University of Agricultural Sciences, Bangalore, India
- R. Shorter, Queensland Department of Primary Industries, Brisbane, Australia

- J.S. Siemonsma, Prosea Project, Centre for Research and Development in Biology, Bogor, Indonesia
- J. Smartt, Department of Biology, The University, Southampton, United Kingdom
- Sumarno, Agency for Agricultural Research and Development, Bogor Research Institute for Food Crops, Indonesia
- E. Westphal, Prosea Project, Department of Tropical Crop Science, Wageningen Agricultural University, the Netherlands

Prosea Project personnel

(January 1989)

Indonesia

M.H. Aarts-van den Bergh, Botany Officer
Ch. Aipassa, Secretary
Jajang, Office Assistant
A.J.G.H. Kostermans, Scientific Adviser
Sampurno Kadarsan, Project leader Indonesia
J.S. Siemonsma, Tropical Agronomist, Regional Coordinator South-East Asia
N.W. Soetjipto, Botanist, Country Officer Indonesia
Hadi Sutarno, Plant Physiologist, Country Officer Indonesia

Malaysia

Salleh Mohd Nor, Project leader Malaysia
Idris Mohd Said, Botanist, Country Officer Malaysia (interim)

the Netherlands

F.J. Breteler, Plant Taxonomist
J.M. Fundter, Forestry Officer
N.R. de Graaf, Silviculturist
J.W. Hildebrand, Forestry Officer
P.C.M. Jansen, Plant Taxonomist
R.H.M.J. Lemmens, Plant Taxonomist
L.P.A. Oyen, Documentalist
J.M.G. Rynja, Secretary
F.M. Stavast, Documentalist until 1 September 1988
E.W.M. Verheij, Horticulturist
E. Westphal, Tropical Agronomist, Project leader the Netherlands
H.C.D. de Wit, Scientific Adviser
W.P.M. Wolters, Project Secretary

Papua New Guinea

R. Banka Adam, Country Officer Papua New Guinea
R.J. Johns, Project leader Papua New Guinea

Philippines

N. Altoveros, Plant Breeder, Country Officer the Philippines
B. del Rosario, Project leader the Philippines

Thailand

Prapandh Boonklinkajorn, Project leader Thailand
Soonthorn Duriyaprapan, Agronomist, Country Officer Thailand

Foreword

In 1986, the Prosea Project published the *Proposal for a handbook* on the plant resources of South-East Asia as a basis for discussion on scientific, organizational and financial aspects of the project, in order to realize the publication of a multivolume handbook. The 'Proposal' was well received, especially by policy makers, specialists and institutions concerned with South-East Asian plant resources.

The first volume of the handbook, on the pulses, is now ready. It is the result of persistant efforts to publish such a handbook. The contribution of an international group of authors, including specialists from South-East Asia, is gratefully acknowledged. In particular, the editors, Professor L.J.G. van der Maesen and Dr Sadikin Somaatmadja, played a major role in structuring the wealth of information into a coherent publication.

As a help to editors and authors, a computerized information system called SAPRIS (South-East Asian Plant Resources Information System), has been established. It contains a checklist of species and information on literature, specialists and organizations. It also contains further information on the species treated in the handbook, as a complement to the printed texts.

The approach taken for this volume is subject to discussion at the Prosea First International Symposium, 22–25 May 1989, at Jakarta, Indonesia. The recommendations of the Symposium will shape future volumes.

This volume, the first in a series, is dedicated to Dr W.M. Otto and Professor Soekiman Atmosoedaryo. Their enthusiasm, wise counsel and continuous support were major ingredients in getting Prosea out of the offices into the reality of South-East Asia.

Wageningen, January 1989

Professor Dr H.C. van der Plas
Rector Magnificus, Wageningen Agricultural University

1 Introduction

1.1 Definition of pulses

This volume deals with pulses of present and future importance in South-East Asia. 'Pulses' are here defined as dry edible seeds of legumes. Legumes are members of the family Leguminosae, bearing pods or legumes. 'Grain legumes' usually mean the same as 'pulses', but leguminous oilseeds are not always included. Summerfield & Roberts (1985) includes soya bean and groundnut in grain-legume crops. FAO statistics keep soya bean and groundnut separate from pulses. In that groundnuts have many confectionary uses and soya beans are eaten as converted pulses (fermented foods and bean sprouts), those legumes are part of this volume too. Some minor pulses, which may have local use, are treated in Chapter 3, as their role is minor. A table (Chapter 4) lists legume species that have another main use and will be treated in other volumes of the Prosea handbook. The seeds of those species are either used locally, or as a snack or as famine food. A cross-reference is provided to the commodity groups where those plants will be described.

1.2 Role

Pulses are of major importance as protein-rich foods. Even when cereals contribute more protein to the human diet in absolute terms as staple food (Zwartz & Hautvast, 1979), cereals are hardly ever eaten without additives to make them palatable and more digestible.

Deficiencies of certain amino acids in pulses, such as the sulphur-bearing amino acids, are at least partly counterbalanced by those in cereals. Biological value of legume proteins is poor but can be improved with methionine (Bressani & Elias, 1980), but their supplementary value to cereal protein is significant. The relative poor digestibility of protein in cooked pulses is a disadvantage that is not well understood, but tannins may play a role. Split dehusked pulses are better digested and more palatable. In combination with cereal grains, amino acids from legumes balance out quite well. So improvement of the amino acid profile of legumes should not be a first priority. Where the diet includes little meat and vegetables, pulses are of prime importance. Religious and economic restrictions often curtail the use of meat, vegetables are often seasonal and cannot be stored long, but pulses (just like cereals) can be kept a long time.

Leguminous seeds are characterized by high contents of protein, up to 40 % (1 % = 10 g/kg) in dry matter, mainly stored in the cotyledons. Non-oilseed pulses are also rich in digestible carbohydrates and provide dietary fibres (Norton et al., 1985). Papilionaceous oilseeds provide major vegetable oils, and the residues contribute protein and small amounts of oil for cattle feed. The embryo and the seed-coat contribute little to the total seed protein, because of their small volume. The seed-coat contains few proteins but considerable amounts of fibrous matter, hence the frequent use of pulses, for instance in India as dhal, split peas, from which the fibrous seed-coats are removed that are used, for instance, as poultry feed.

Table 1 compiles the average composition of legume seeds, taken from the data provided by the contributors to this volume. Soya beans and winged beans score highest for protein, and winged beans have about the best content of cystine and threonine. Winged beans are promising as dry grain, though they are not yet used as a pulse but as a vegetable. The limiting amino acids do not widely differ between the pulses, and are compensated in balanced diets by cereal amino acids. The discrepancies between these data and those presented elsewhere (e.g. Norton et al., 1985) illustrate the point that genotype–environment and season interactions influence the seed composition. No single source pro-

Table 1. Composition of legume seeds (per 100 g edible portion) as mass fraction (g/100 g) or for limiting amino acid in total nitrogen (mg/g N) and energy value.

Crops	Mass fraction						Limiting amino acids					Energy (kJ)
	water	protein	fat	carbohydrates	fibre	ash	Met	Cys	Thr	Try	Val	
Arachis hypogaea	5.4	30.4	47.7	11.7	2.5	2.3	70	80	160	70	260	2457
Cajanus cajan	10	20	2	59	7.2	3.8	80	70	180	30	230	1450
Cicer arietinum	9.9	20.6	5.7	57.8	3	3	80	90	240	50	280	1520
Glycine max	10	35	18	32	4	5	80	100	240	80	300	1680
Lablab purpureus	9.6	24.9	0.8	60.1	1.4	3.2	56	·	263	·	325	1403
Lathyrus sativus	10	25	1	61	1.5	3	24	·	150	25	290	1231
Lens culinaris	11	24	1.8	59	1.8	2.2	50	60	250	60	310	1450
Macrotyloma uniflorum	10	22	0.5	61	5	3	106	62	·	60	356	1400
Phaseolus acutifolius	10	24	1.5	56.5	4	4	60	90	250	·	360	1588
Phaseolus coccineus	12.5	20.3	1.8	62	4.8	3.4	·	·	·	·	·	1420
Phaseolus lunatus	13.2	20	1.5	58.2	3.7	3.4	90	90	260	60	320	1450
Phaseolus vulgaris	10	22.6	1.4	62	4.3	3.7	70	50	250	60	290	1453
Pisum sativum	10	25	1	57	4.5	2.5	60	70	250	60	290	1428
Psophocarpus tetragonolobus	9.7	32.8	17	32	5	3.5	75	100	270	60	300	1697
Vicia faba	10	26	1	59	7	3	44	50	210	·	280	1450
Vigna aconitifolia	10.8	23.6	1.1	56.5	4.5	3.5	·	·	·	·	·	·
Vigna angularis	10.8	19.9	0.6	64.4	7.8	4.3	110	70	240	·	340	1411
Vigna mungo	10	22	1	62.9	0.9	3.2	90	60	230	·	370	1455
Vigna radiata	10	22	1	60	4	3	30	40	210	·	260	1430
Vigna subterranea	11	18	6	62	5	3	113	70	219	·	340	1540
Vigna umbellata	13.3	20.9	0.9	64.9	4.8	4.2	769	44	294	·	394	1373
Vigna unguiculata	10	22	1.4	59.1	3.7	3.7	75	70	225	70	280	1420

Sources: contributors to this volume and various other references (values are sometimes difficult to find, to compare and contradictory).
Met = Methionine, Cys = Cystine, Thr = Threonine, Try = Tryptophane, Val = Valine.
· = data not available.

vides comparable data for all pulse species and ranges for all contents would be more appropriate.

In 1980–1987, the average world production of the non-oilseed pulses reported by FAO (1988) amounted to 39 382 000 t/year (1 tonne = 1000 kg) from an estimated 56 701 000 ha (1 hectare = 10 000 m^2 = 0.01 km^2). Total production of soya bean and groundnut was 90 603 184 and 18 469 728 t/year from 51 448 816 and 18 148 480 ha, respectively. In South-East Asia (FAO, 1988) the 8 reporting countries), yields of pulses were 1 476 000 t/year from 1 986 122 ha (8-year average 1980-1987), but at least 15 % from small production and kitchen gardens is not reported. Values are listed in Table 2. Production of soya beans and groundnuts in shell were 1 104 000 and 1 723 000 t/year from 1 179 000 and 1 436 046 ha, respec-

Table 2. Average area and production of seed legumes in South-East Asia (1980–1987) (Source: FAO, 1988).

	Burma	Cambodia	Indonesia	Laos	Malaysia	Philippines	Thailand	Vietnam	South-East Asia	World
Area (ha × 1000)										
Dry beans	435	38	403	.	.	48	462	51	1 437	25 959
Dry peas	27	.	.	.	.	.	.	.	27	8 428
Chick peas	170	.	.	.	.	–	.	.	170	9 729
Cowpeas	23	.	.	.	.	6	.	.	29	3 766
Pigeon peas	60	.	.	.	.	–	.	.	60	3 408
Other pulses[1]	9	.	7	11	.	–	105	131	263	5 411
Soya beans	27	1	849	6	0[2]	9	184	103	1 179	51 449
Groundnuts	539	8	516	12	5	50	118	188	1 436	18 148
Total										
excl. oilseeds	724	38	410	11	.	54	567	182	1 986	56 701
incl. oilseeds	1 290	47	1 775	29	5	113	869	473	4 601	126 298
Production (t × 1000 per year)										
Dry beans	306	30	339	.	.	35	299	73	1 082	14 637
Dry peas	22	.	.	.	.	.	.	.	22	11 361
Chick peas	142	.	.	.	.	.	.	.	142	6 396
Cowpeas	10	.	.	.	.	5	.	.	15	1 345
Pigeon peas	36	.	.	.	.	.	.	.	36	2 471
Other pulses[1]	5	.	4	22	.	.	73	75	179	3 172
Soya beans	20	2	782	5	0[2]	8	216	71	1 104	90 603
Groundnuts	531	8	775	10	20	43	155	181	1 723	18 470
Total										
excl. oilseeds	521	30	343	22	.	40	372	148	1 476	39 382
incl. oilseeds	1 072	40	1 900	37	20	91	743	400	4 303	148 455

1. Other pulses (1980–1986 figure) include *Lablab, Lathyrus, Lens* (only reported from Burma), *Macrotyloma, Vigna subterranea, Trigonella foenum-graecum* etc.
2. Only 39 ha, resp. 58 t.

tively. With 937 and 1199 kg/ha, the oilseeds in the area outyield the pulses with their average yield of 742 kg/ha, which leaves much to be desired. The 8-year average statistics of production in South-East Asia are below the 1986 and 1987 values. The economically more valuable soya bean and groundnut (cash crops) are presumably grown on better soils with better care. The minor pulses are not reported in the statistics, nor do they find their way to the statistical yearbooks.

Pulses have not witnessed spectacular growth worldwide. Over the 1970s, the trends were stable or production increased (soya bean and groundnut) or area increased (beans, peas) with the occasional good or bad year. In certain countries, some crops have gained ground, as have chickpeas in Turkey. Pulses of dry areas fluctuate more than those of wetter (and more stable) climates. In South-East Asia, soya beans are on the increase in Indonesia, though with lower estimates for 1988. Groundnuts are stable on the whole but had good years in 1986 and 1987, with lower estimates for 1988 in Indonesia. Areas under soya bean and production of soya bean in the Philippines have dropped from a maximum in 1982.

International trade in pulses is increasing in volume and value. In South-East Asia, net exporters (1983–1985) are Burma and Vietnam; net importers are Indonesia, Malaysia and the Philippines. Malaysia exports more soya oil than it imports; the other countries are net importers of small to considerable amounts. Groundnut oil is hardly traded between countries at all in the Prosea region, apart from the large amounts transferred through Singapore, with alternate export and import surpluses. Local trade is large.

On a world scale, cereals occupy more than ten times as much arable area as non-oilseed legumes and produce 35 times as much raw product, illustrating the need to improve pulse productivity.

1.3 Botany

1.3.1 Taxonomy and wild relatives

The major edible pulses belong to the subfamily Papilionoideae, being the largest subfamily of the Leguminosae with 450 genera and 10 000 species. The subfamilies Caesalpinoideae (180 genera, 2700 species) and Mimosoideae (55 genera, 2000 species) include some condiments and emergency foods; otherwise those subfamilies are more useful for timber, fuel, medicine and various other purposes. The subfamily Papilionoideae includes also those legumes most frequently used for fodder and green manure. Caesalpinoideae and Mimosoideae are predominantly tropical.

Keys to the genera of pulses on a world scale are given by the contributors to Polhill & Raven (1981). Keys to the genera and species are available in local floras such as Backer & Bakhuizen-van den Brink (1963), Nguyên Van Thuân (1979) and Verdcourt (1979). The Leguminosae from large areas in South-East Asia are not yet covered by modern floras.

Only a few tribes of legumes contribute the main pulses. The Vicieae are more relevant outside the tropics and includes the genera *Lathyrus*, *Lens*, *Pisum* and *Vicia*. *Cicer* is now classified in a tribe of its own, Cicereae. The tribe Phaseoleae includes the most species of pulse, including the many species of *Phaseolus*,

Vigna, Cajanus and *Glycine*. Table 3 lists the taxonomic arrangement of the major pulses and legumes with edible seeds as secondary use in South-East Asia and indicates the existence of wild relatives, a major source of material if conventional sources of breeding stock are inadequate.

The wide range of types created by man and nature over centuries during selection of cultivated plants have been variously classified. The most recent approach is to distinguish cultivars and to group these in cultivar groups. That approach (to be followed throughout the volumes of Prosea) has not been formally adopted for all legumes treated here but is preferable. In the literature, closely related legume species are often reduced to subspecies or varieties (the formal classification under the jurisdiction or the International Code of Botanical Nomenclature). Although that may be useful, classification into cultivars and cultivar groups (the informal 'open' classification guided by the International Code of Cultivated Plants) is advisable. Practical and regional uses of cultivar groups may only provide partial solutions to infraspecific classifications but these prove workable in various conditions.

Without following the use by Harlan & De Wet (1971) of the category subspecies, their approach of discerning primary, secondary and tertiary genepools is useful. The primary genepool consists of cultivated and wild forms of the same biological species. The secondary genepool consists of related species that hybridize with the cultigen with some difficulty, and whose hybrids are at least partly fertile. Species classified in the tertiary genepool can be crossed with the cultigen only by advanced techniques, and their hybrids are usually sterile, anomalous or even lethal. The boundaries differ according to results from hybridization and interpretation. For instance, *Cicer reticulatum* Ladiszinsky crosses relatively easily with *Cicer arietinum* L., and can be considered either a variety of the latter or a species in the secondary genepool; the crossing barrier is not strong. Other related annual relatives of chickpea belong in the secondary or tertiary genepool; related perennials cannot be placed in a genepool of the chickpea yet.

Particularly in the breeding of groundnut, the usual sources of disease resistance have not been ample and wild *Arachis* species constitute parental material. There is also some promise in the wild germplasm of *Cajanus* spp. (earlier known as *Atylosia* spp.), *Vigna* spp. and *Phaseolus* spp. However no species akin to faba bean, *Vicia faba* L., have ever crossed with the cultigen. *Vigna* is a large genus of 150 species, mainly in Africa and Asia, now at last well separated from *Phaseolus*, whose 50 species all originate in the Americas. *Vigna unguiculata* (L.) Walp. comprises three cultivar groups: Unguiculata, Biflora and Sesquipedalis, and three wild subspecies: *dekindtiana* (Harms) Verdc., *tenuis* (E. Mey.) Maréchal, Mascherpa & Stainier and *stenophylla* (Harvey) M., M. & S. Subspecies *dekindtiana* includes the botanical varieties *dekindtiana*, *mensensis* (Schweinf.) M., M. & S., *pubescens* (Wilczek) M., M. & S., and *protracta* (E. Mey.) Verdc. Steele et al. (1985) seem to consider the wild subspecies as secondary genepool, to which *Vigna nervosa* Markötter may belong, as it is part of Section *Catiang* (DC.) Verdc. of Subgenus *Vigna* too. Other species of *Vigna* do not cross with cowpea. Mung bean and black gram belong to *Vigna* Subgenus *Ceratotropis* (Piper) Verdc., just like *V. aconitifolia* (Jacq.) Maréchal, *V. angularis* (Willd.) Ohwi & Ohashi and *V. umbellata* (Thunb.) Ohwi & Ohashi. Subgenus *Ceratotropis* comprises 18 species in the treatment of Verdcourt (1970)

Table 3. Alphabetical arrangement of genera of some pulses and legumes with edible seeds as secondary use, subfamily, tribe and subtribe, the number of species per genus and availability of a monograph.

Adenanthera L. (Mimosoideae): Mimoseae Bronn, Adenanthera group. Ca. 8 species in tropical Asia and the Pacific. (−)
Arachis L. (Papilionoideae): Aeschynomenae (Benth.) Hutch., Stylosanthinae (Benth.) Rudd. Ca. 60 species in South America. (−)
Caesalpinia L. (Caesalpinioideae): Caesalpinieae, Caesalpinia group. Ca. 10 species pantropical. (±)
Cajanus DC. (Papilionoideae): Phaseoleae DC., Cajaninae Benth. 32 species in South, South-East and East Asia. (+)
Callerya Endl. (Papilionoideae): Milletieae Miq. 16 species in South-East Asia, northern Australia and Japan. (−)
Canavalia DC. (Papilionoideae): Phaseoleae DC., Diocleinae Benth. Ca. 50 species in the tropics, esp. America. (+)
Cassia L.[1] (Caesalpinioideae): Cassieae Bronn, Cassiinae. Ca. 30 species in the tropics. (+)
Chamaecrista[1] (Caesalpinioideae): Cassieae Bronn, Cassiinae. 250 species in the neotropics and some elswhere. (±)
Cicer L. (Papilionoideae): Cicereae Alef. 43 species in Central Asia, south-western Asia and the Mediterranean. (+)
Crotalaria L. (Papilionoideae): Crotalarieae (Benth.) Hutch. Ca. 600 species in the tropics, esp. Africa. (+)
Cyamopsis DC. (Papilionoideae): Indigoferae (Benth.) Rydb. 3 species in Arabia and Africa. (±)
Dolichos L. (Papilionoideae): Phaseoleae DC., Phaseolinae Benth. 60 species in Africa to East Asia. (+)
Entada Adans. (Mimosoideae): Mimoseae Bronn, Entada group. 30 species in the tropics, esp. Africa. (±)
Glycine Willd. (Papilionoideae): Phaseoleae DC., Glycininae Benth. 9 species in East Asia and Australia. (±)
Inocarpus J.R. & G. Forster (Papilionoideae): Dalbergieae Bronn ex DC. 3 species in Malaysia, Indonesia, Philippines, Papua New Guinea and the Pacific. (−)
Intsia Thouars (Caesalpinioideae): Detarieae DC., Hymenostegia group. 3 species in Asia and the Pacific. (−)
Lablab Adans. (Papilionoideae): Phaseoleae DC., Phaseolinae Benth. 1 species in Africa. (+)
Lathyrus L. (Papilionoideae): Vicieae (Adans.) DC. Ca. 150 species in Europe, Asia and North America. (+)
Lens Mill. (Papilionoideae): Vicieae (Adans.) DC. 5 species in the Mediterranean and Ethiopia. (±)
Leucaena Benth. (Mimosoideae): Mimoseae Bronn, Leucaena group. 40 species in Central and South America. (−)
Lysidice Hance (Caesalpinioideae): Detarieae DC., Hymenostegia group. 1 species in Vietnam and southern China. (+)
Macrotyloma (W. & A.) Verdc. (Papilionoideae): Phaseoleae DC., Phaseolinae Benth. 24 species in tropical Africa and Asia. (+)
Mucuna Adans. (Papilionoideae): Phaseoleae DC., Erythrininae Benth. 100 species in the tropics and subtropics. (−)
Padbruggea Miq. (Papilionoideae): Millettieae Hutch. Ca. 10 species in South-East Asia. (−)
Parkia R. Br. (Mimosoideae): Parkieae (W. & A.) Benth. 40 species in the tropics. (−)
Phaseolus L. (Papilionoideae): Phaseoleae DC., Phaseolinae Benth. Ca. 50 species in tropical and subtropical America. (+)
Pisum L. (Papilionoideae): Vicieae (Adans.) DC. 2 species in the Mediterranean. (+)
Pithecellobium (Mimosoideae): Ingeae Benth. 20 species in Central and South America. (−)
Psophocarpus DC. (Papilionoideae): Phaseoleae DC., Phaseolinae Benth. 9 species in the paleotropics. (+)
Senna Mill.[1] (Caesalpinioideae): Cassieae Bronn, Cassiinae. 240 species in the tropics. (+)
Sesbania Scop. (Papilionoideae): Robinieae (Benth.) Hutch. 50 species in the tropics and subtropics. (−)
Shuteria W. & A. (Papilionoideae): Phaseoleae D.C., Glycininae Benth. 5 species in South-East Asia. (+)

Tamarindus L. (Caesalpinioideae): Amherstieae Benth. emend. Leon, Amherstia group. 1 species in tropical Africa. (+)
Trigonella L. (Papilionoideae): Trifolieae (Bronn) Benth. 80 species in the Mediterranean, Central Asia and Australia. (+)
Vicia L. (Papilionoideae): Vicieae (Adans.) DC. Ca. 140 species in Europe, Asia and North America. (−)
Vigna Savi (Papilionoideae): Phaseoleae DC., Phaseolinae Benth. Ca. 150 species in Africa and Asia. (+)

1. *Chamaecrista* and *Senna*, formerly subgenera of the genus *Cassia*, are now recognized as genera, leaving *Cassia* (in the narrow sense) a small genus with only 30 species.
(+) = monograph available, (−) = monograph not available, (±) = partial or floristic treatment of more than 1 or 2 species available.

and some are cross-compatible with others. Each cultivated member of the groups can be crossed successfully with at least one other but hybrids are often sterile.
A secondary and tertiary genepool of *Glycine* was considered not to exist (Harlan & De Wet, 1971), but hybrids have been made with great pains between *Glycine max* (L.) Merr. and *G. tomentella* Hayata by culture in vitro. Recently two other species, *G. argyrea* Tindale and *G. cyrtoloba* Tindale have been discovered in Queensland.
Hybridization between genera is difficult in Leguminosae (McComb, 1975) and, if successful, provides a criterion to reconsider generic boundaries. Proper taxonomic classification and nomenclature are the keys to exchange of information. For many genera, modern monographic or floristic treatments are available (bibliographies; Polhill & Raven, 1981; Polhill & van der Maesen, 1985). Infraspecific variation is more complicated. For more about advances in legume systematics, see Stirton (1987).

1.3.2 Morphology

Most of the cultivated pulses are papilionoid annuals or are grown as annuals. Many wild relatives have a perennial life-cycle, usually of a couple of years (*Cajanus* spp.). Growth habit is either climbing or trailing, bushy or intermediate (bushy with trailing stems). The climbers are indeterminate and the bushes determinate; habit is also related to flowering. The stems produce alternate, trifoliolate or pinnate leaves. Inflorescences are (pseudo)racemes of 1 to many flowers, and the many-shaped pods originate from simple carpels. Seeds have two cotyledons storing the food reserves man uses.
Leaves are simply or double compound; even single-leaflet leaves bear pulvini at the base of the petiole, allowing the typical sleeping position (nyctinasty) that many legumes show at night, and allowing movement during the day. The petiolules of the leaflets may be thickened and act as pulvinus.
Papilionoid flowers follow the basic zygomorphic pattern of legumes with 5 sepals, 5 petals, superior ovary, and stamens a multiple of 5. The calyx is usually merged into a campanulate structure with 5 teeth, the upper ones often connate, and distinct petals; the flag, standard or vexillum is the upper, exterior, usually largest petal, attracting insects; the wings or alae are lateral and cover the

keel petals or carina, which covers stamens and ovary. Caesalpinoid flowers are also zygomorphic and quite butterfly-like, but the upper largest petal is covered by the lateral petals; the lower petal and the calyx parts are usually free, the stamens are free and often reduced in number. Mimosoid flowers are actinomorphic (radial symmetry) and have 5–10 or many conspicuous stamens. United in globose heads or (dense) spikes, the otherwise small flowers appeal to insects by providing aromatic nectar and pollen. In the pulses under consideration, the stamens are monadelphous (10 together) or diadelphous (9 together, the upper or vexillary stamen free).

The root system consists of a well developed tap-root with lateral secondary roots, usually in 3 vertical rows (Vicieae and Phaseoleae), or 4 rows, sometimes more. The seedlings emerge with the cotyledons either buried (hypogeal) or exposed above ground (epigeal). The first leaves are small scales (cataphylls), or simple or reduced leaves (eophylls), the latter, for instance in beans, opposite, before leaves of typical shape are produced.

Fruits (legumes) differ in shape, varying from woody underground pods (*Arachis*, *Vigna subterranea* (L.) Verdc.) to succulent vegetable pods 1 m long (*Vigna unguiculata*), at least in the young edible stage, to winged or inflated structures with or without hairs. Seeds vary in size and shape, usually with a shiny or dull seed-coat of various colours, which may fade in storage. The place of attachment to the funiculus, the hilum, varies from circular to long-extended (*Lablab purpureus* (L.) Sweet) and may be provided with a rim-aril or strophiole.

1.3.3 Growth and development

Root systems depend on the nature of the crop. Drought-tolerant species (*Cicer*, *Cajanus*, *Lens*, *Macrotyloma*) have deeply penetrating roots (1–2 m). Legumes need permeable well drained soil with good aeration. The tap-root forms lateral secondary roots in three vertical rows (triarch), and the upper 15 cm of soil usually contains 80–90 % of the total dry weight of roots. Long vertical roots search further down for soil water.

All cultivated pulses benefit from symbiosis with bacteria (*Rhizobium* spp.), which induce the roots to produce nodules, providing shelter for the bacteria to multiply and fix nitrogen gas from the air, providing fixed N to the plant and supplying carbohydrates to the bacteria. Interaction between legumes and *Rhizobium* species and strains is often specific, with native populations of the bacterium in the soil forming a major factor. For details on formation of nodules, see Cobley & Steele (1976), Pate & Minchin (1980) and Sprent & Minchin (1985). Classification of the *Rhizobium* species infecting cultivated legumes, static since the 1930s, is now under review. The shape of the root nodules is often characteristic (Corby, 1981) and has systematic value. Root-nodule bacteria have coevolved with legume species and populations. Legume nodules may fix nitrogen to the soil even up to 100–200 kg per ha per season, of which 70 % is directed to the pods. Much of the residue of aerial parts is used and removed from the field.

A major characteristic in growth habit and flowering is the distinction between *determinate* and *indeterminate* cultivars. Legume species often have both forms: the bushy and climbing shapes. Intermediate forms also exist. Determinate flowering is defined as having flowers limited in numbers, indeterminate flowering produces many flowers over a longer period of time. Those terms should not be confused with 'predominantly terminal' (basipetalous) and 'mainly axillary' (acropetalous), as used by Rachie & Roberts (1974; see also van der Maesen, 1986), that is with determinate or cymose inflorescences, and indeterminate or racemose inflorescences, respectively. Determinate inflorescences start with the terminal flower; in indeterminate inflorescences, the lowest periferal flower opens first. In pigeon pea, the inflorescences are acropetalous pseudoracemes, whether the plant is determinate or indeterminate. Determinate cultivars produce their axillary and terminal pseudoracemes at about one level and flowering is soon over. Indeterminate cultivars produce inflorescences at more nodes, while the plant is growing, producing flowers all along the branches; flowering may last several months. That is advantageous if early flowers are damaged, for instance by insects or bad weather.

Phaseoleae may have panicles, pseudoracemes or pseudoracemes with nodes (Lackey, 1981). *Cicer* and Vicieae have solitary flowers or axillary racemes, rarely panicles, Aeschynomeneae (*Arachis*) species bear racemes or panicles. Determinate cultivars do not increase much in height after flowering starts ('bush types') and flower briefly; indeterminate forms ('climbing types') continue to grow, reach 2–4 times the height of the bush types and flower over a long period. Indeterminate cultivars can be considered more primitive, that is less adapted to cultivation and more adapted to compensate, for instance, for insect attacks. In the quest for high-yielding short-duration cultivars, determinate forms are usually practical but may need a pesticide umbrella. In mixed intercropping, climbing or trailing forms are attractive and productive, for instance with tall cereals for support.

Flowering may depend upon (seed) vernalization, though cultivars adapted to the tropics hardly react at all to cold. Flowering is influenced by daylength and temperature. Some species, particularly those originating from higher latitudes are quantitative long-day plants (chickpeas, lentils); many other species flower sooner after induction by short days (soya beans, common beans, Lima beans, pigeon peas). Rarely do cultivars not flower at all under the wrong daylength but delay in flowering reduces productivity. However the usual cultivation of pigeon pea in South Asia and parts of Africa coincides with the rainy long-day season, providing 3–4 months of vegetative growth before flower induction when days get shorter. With short days, pigeon peas flower after 30–50 days, allowing high plant populations and short-duration cropping. Many legume cultivars are day-neutral. Interaction with temperature is often complicated and locally adapted cultivars, for instance of common beans and cowpeas, fit into traditional cropping systems. Growth habit is variable and also depends on environment.

Flower formation and fertilization may be unfavourably affected by warm or cool weather (more than 30 °C or less than 15 °C), low atmospheric humidity, water stress and incorrect daylength. Papilionoid flowers attract many insects, irrespective of the nature of the pollination in pulse species, which can be completely self-pollinated to mainly cross-pollinated. In most pulses, pollen from the flower's own stamens are most likely to reach the stigma. Chickpeas, soya beans, common beans, Lima beans and *Phaseolus aconitifolius* are strictly self-pollinated and self-fertilized; peas usually have less than 1 % outcrossing; *Phaseolus coccineus* is normally cross-fertilized; *P. lunatus* ranges from complete

Table 4. Pollination as proportion cross-fertilized (%) in some pulses under natural conditions.

Arachis hypogaea	0–2.56 (–6.26)
Cajanus cajan	3–45(–65), 15–20 is common
Cicer arietinum	0
Glycine max	0
Lablab purpureus	frequent
Lathyrus sativus	likely self-pollinated
Lens culinaris	0 (–some?)
Leucaena leucocephala	usually self-pollinated
Macrotyloma uniflorum	likely self-pollinated
Phaseolus acutifolius	0 (–some?)
Phaseolus coccineus	100, probably 30–40 is normal
Phaseolus lunatus	0–100, up to 20 also recorded
Phaseolus vulgaris	0 (–1–10)
Pisum sativum	0–1
Vicia faba	25–50
Vigna aconitifolia	normally 0
Vigna angularis	frequent
Vigna mungo	0
Vigna radiata	hardly any
Vigna subterranea	low
Vigna umbellata	some
Vigna unguiculata	very low < 1%, to considerable (humid United States, Nigeria)

Compiled from various sources.

self-fertilization to cross-fertilization depending on cultivar, plant spacing, prevailing wind direction and insect populations. Pigeon peas are cross-fertilized from 3 to 50 % or more (Table 4).

The term cleistogamy (pollination in closed flowers) is often mistakenly applied to papilionaceous flowers. Although many species shed pollen and the pistil grows through the anther tube sweeping up all the pollen hidden in the unopened flower bud, most flowers open later and attract pollinators. Cold weather may cause the flowers not to unfold the flag. Colours and nectar attract insects. In some species, certation may play a role: foreign pollen tubes may reach the ovules sooner than pollen tubes from anthers of the same flower. Cross-pollination followed by cross-fertilization is a problem in producing certified seed, requiring spatial isolation of 100–200 m. Most pulses, apart from runner beans, may produce seeds in the absence of pollinators but often set more seed if pollinated by insects such as bees or humble-bees. The role of thrips in pollination is negligible; cross-fertilization by thrips is extremely unlikely and has not been proved. Many flowers abort in some species (Summerfield & Wien, 1980), especially in the more primitive (woody) pigeon pea.

Tucker (1987) explored flower initiation and anatomical development. Some ontogenetic features characterize the subfamilies; others (increase or decrease of parts, fusion, loss of parts, form changes) characterize tribes. Loss of petals and sepals is rare; suppression after initiation is common. The stamens in all legumes are free at first; fusion causes monadelphy or diadelphy. Knowledge of sterility in stamens, as it occurs in several stages of development, may lead to production of male-sterile genotypes. Inflorescence structure depends on suppression of certain floral apices.

Fruit and seed development have been studied in a few pulses. The seeds attain their maximum dry weight between 30 and 70 days after anthesis. Pate et al. (1988) and Herridge et al. (1988) provide details on partitioning of carbon and nitrogen in the nodulated grain legume. Nitrogen is rerouted from its initial direction towards lower leaves; the xylem stream becomes more concentrated in nitrogen as it moves upwards to benefit apically sited leaves and flowers. Much of the plant's need for nitrogen comes from nodule activity during vegetative growth. Modes of translocation and fixation of nitrogen differ between species: in soya bean and cowpea, ureides are the nitrogen compounds used; in pea and other cool-season legumes, amides (asparagine and glutamine) are the vehicle of nitrogen transport.

A typical annual pulse produces the largest seeds and pods from the first well developing and fertilized flowers. The first flower buds may not develop, for instance 'pseudoflowers' in chickpea. Later, when roots and leaves senesce, the source limits full transfer to the sink, i.e. pods are less filled and seeds tend to have a lower weight. Some perennial legumes (pigeon pea) seem to have a more uniform seed weight, whether produced early or late in season. However at a certain point, even those indeterminate legumes cease flowering and pod setting when the sink has reached a sustainable level. Genotypes differ in photosynthetic activity; some genotypes keep leaves green longer than in other ones.

1.4 Ecology

Most pulses in South-East Asia belong to the (sub)humid lowland tropics. Conditions similar to temperate climates at higher altitudes produce (vegetable) legumes like those outside the tropics. The semiarid climates of low and intermediate altitudes, usually found at higher latitudes, provide a home for short-duration and drought-resistant cultivars. Several species of pulses include variously adapted cultivars.

1.4.1 Climates

Rachie & Roberts (1974) produced the following useful classification, discerning between low and medium–high altitudes.

- Tropical lowland, semiarid: altitude 0–800 m, < 600 mm rainfall; short-duration groundnuts, cowpeas, pigeon peas; less important are Bambara & Kersting groundnuts, moth beans, cluster beans, horse gram and tepary beans.
- Tropical lowland, subhumid: altitude 0–800 m, 600–1000 mm rainfall; medium and long duration groundnuts, cowpeas, pigeon peas, mung beans, black gram; secondary pulses are lablab beans and horse gram; when rain is well distributed, this zone produces common beans, soya beans, rice beans and Jack beans too.
- Tropical lowland, humid: altitude 0–800 m, 1000–1500 mm rainfall; medium and long-duration cowpeas, pigeon peas, mung beans, black gram dominate; further groundnuts, Lima beans, common beans, rice beans, Jack beans and sword beans are found.
- Tropical lowland, humid: altitude 0–800 m, > 1500 mm rainfall; African and Mexican yam beans, Lima beans, velvet beans and winged beans are impor-

tant; other species have adapted cultivars.
- Medium and high altitudes in the tropics: altitude > 800 m; soya beans, common beans, chickpeas and peas are dominant (chickpeas require a dry season; the other pulses prefer dry weather when ripening) and less important are groundnuts, cowpeas, pigeon peas, runner beans, lentils, horse beans and grass peas; lentils, grass peas and chickpeas are typically adapted to cooler seasons after the monsoon at higher latitudes and the subtropics, also at lower altitudes.

Temperature regimes vary between species and sometimes cultivars differ in adaptation. The temperate crops such as common beans and peas withstand lower temperature (no night frost). Semiarid crops such as chickpeas, lentils and grass peas may survive occasional chills in early stages. Chickpea seedlings are even known to revive from snow cover. The drought-resistant species withstand the largest temperature amplitudes.

Bright sunshine is essential for growth and fertilization of semiarid pulses. Cloudy weather harms fruit setting in chickpeas. Long-duration climbers such as yam beans and velvet beans may tolerate shade and moist conditions.

1.4.2 Soils

Pulses grow well on well aerated drained soils, of medium fertility, sustaining *Rhizobium* bacteria and other symbiotic organisms (mycorrhiza). Pulses have been the major reason why agriculture was sustainable for centuries without introducing large amounts of fertilizer or manure. Rotation and intercropping with legumes is an old tradition in many areas.

Drought resistance is often based on deep rooting systems, permitting the plant to tap residual moisture deeper in the soil. Drought is also tolerated because of special features, such as movable leaflets, closing of stomata, glandular secretion (e.g. chickpea) and dense indumentum (i.e. hairs). Soil pH is preferably between 6 and 7.5; semiarid pulses are known to grow at pH above 8. Acid soils may sustain Kersting and Bambara groundnuts, winged beans and yam beans.

Despite many efforts, few cultivars, if any, adapted to alkaline or saline soils have been found. Areas with such soils should be planted with cereals resistant to such conditions.

1.5 Agronomy

1.5.1 Place of pulses in cropping systems

Pulses have not undergone such a relative increase in area as cereals, despite the much higher price they fetch. Earlier in the 20th Century, prices of pulses and cereals were similar in the area. Because of the inherent uncertainty from the more erratic and lower yields, legume species retain a secondary place, though suitable for the rotation and for intercropping systems. The attitude of the farmers to risks prevents large changes in the area, as cereals are more certain as a crop. Still, the nitrogen-fixing bacteria are necessary for other crops following or intercropped with legume species.

Mixed (inter)cropping is a frequent practice. Its use in small cropping systems is well known. The use of mixtures reduces the feasibility of mechanization, for instance of intercultivation, pesticide application and harvesting. Nevertheless, mixed intercropping should often be regarded as a highly adapted practice rather than an out-of-date one. Small home gardens, as in Indonesia, typically have multiple crop mixtures of more than two species. In semiarid regions, wheat and chickpeas in mixtures ensure reasonable harvests with more cereals, if rains are above average, and more pulses, if little rain falls. Roots of pulses penetrate deeper than those of cereals; in combination, they can exploit the soil more effectively for water and nutrients. Maize–*Phaseolus*, wheat–chickpeas, sorghum–pigeon peas are just some of the best known crop mixtures. Usually intercropping is with single or multiple rows (to 10–20 rows) of one crop to the other. Many situations between row intercropping and strip intercropping are known. Relay sowing is practicable but not (yet) popular; chickpeas, soya beans and mung beans can be broadcast in standing rice crops during drying out. That is a zero-tillage system.

Green-manure species have special use. Small-seeded legume species or pulse cultivars are preferable, as their seed harvests compete less with food grains; the seed rates are low, 5–10 kg/ha. However when decisions have to be made whether to improve the soil with green manures, farmers usually decide to direct inputs in labour, water, space and time to edible (cash) crops. The residues of pulses are valuable as fodder or fuel, and, in subsistence farming, the residues are rarely returned to the soil.

1.5.2 Sowing practices

Pulses are grown from seed. Vegetative propagation is sometimes feasible, usually for research purposes or when species are tuber-bearing (winged beans, yam beans, some *Phaseolus* spp.). Seeds may survive deep sowing, which is useful under hot dry conditions in light soils, to protect early germination by light showers. Sowing time is as close as possible to the onset of the first rains for the long-duration monsoon crops. Pulses after the monsoon should be sown as soon as rains cease, to avoid drying of the seed-bed; tillage should disturb the upper layers as little as possible. Larger-seeded legumes are best placed between 2 and 5 cm deep. Seed rates are high for large-seeded cultivars and species, up to 120 kg, but, in subsistence farming, such rates are rarely used.

Seed-bed preparation follows that of the main crop and often does not require fine tillage, as pulse seeds are larger than cereal grains. Proper aeration is required.

Seed of pulses should often be treated with a seed dressing, for instance a combination of a fungicide and an insecticide. However several fungicides (particularly those containing heavy metals such as mercury), many pesticides and some herbicides are not only toxic to *Rhizobium* cultures but also reduce nodulation in the field. If seed treated with toxic agrochemicals must be inoculated, the rhizobia must be spatially separated either by direct inoculation into the soil or onto the seeds previously coated with a poly(vinyl acetate). In practice, few pulse seeds are either inoculated or dressed. Inoculation is a paying proposition if the soil has not been cultivated with the pulse crop for many years.

1.5.3 Management

Ideally pulse crops should be properly managed by weeding, some irrigation and by fertilizers, but usually crop care is kept to a minimum. Just as cereals compensate by tillering, legume species can compensate for gaps due, for instance, to pests or failure to germinate by branching. Indeterminate cultivars are better equipped to do so, even flowering may be prolonged, if initial flowers are attacked by insects. Staking is required for climbing species of bean, unless support cereals (maize, sorghum) perform that duty. Fences and hedges (e.g. pigeon peas) may be used to support climbers. Irrigation is applied in (semi)arid areas, if residual moisture is insufficient. Otherwise pulses are rarely irrigated. Fertilizers are generally uneconomical for pulses. Nutrient uptake, in combination with nitrogen fixation, is often sufficient from the soil (Summerfield & Bunting, 1980). Pulses are efficient burrowers and, as such, also help the next crop in rotation. Return of plant residues is commendable to improve soil fertility and structure, but most legumes have straw palatable for cattle and this by-product is valuable. In some fuel-scarce areas, the stems of tall cultivars of pigeon pea fetch more money than the seeds they produce. Groundnuts and soya beans, which have probably the most advanced cultivars, receive light dressings of starter nitrogen, 20–30 kg/ha. A good start usually proves economical.

Diseases and pests, both specific and polyphagous ones, harm pulse crops and stored pulses. Chemical treatment is feasible but is often uneconomical. With ultralow-volume (ULV) spraying, even the tall crops can be treated. Advances have been made in producing resistant cultivars but much research is needed to see whether insects still avoid resistant lines, if they have no alternative.

Of the many hundreds of insect species attacking legumes and several dozens of diseases, a few attack most pulses. Disease agents such as fungi and viruses are often specific to certain crops or have a narrow host range. However insect pests are commonly polyphagous, but narrow adaptation to the host exists too. Without thorough calculations, the following pests are probably the most serious: *Heliothis zeae*, *H. armigera*, bollworms, pod-borers, fruitworms or gram caterpillars; *Exelastis atomosa*, plume moths; *Maruca testulalis*, spotted pod-borers; *Melanagromyza phaseoli*, beanfly; *Callosobruchus chinensis*, bruchid beetles or bean weevils. Other beetles, aphids, thrips, leafhoppers and spider mites are also pests of pulses.

The most serious diseases are probably leaf-spots or blight diseases, *Ascochyta*, *Colletotrichum*, *Cercospora*; mosaic viruses: bean yellow mosaic virus (BYMV), cowpea aphid-borne mosaic virus (CABMV), rosette; root-wilts, *Fusarium*, *Sclerotium*, *Pseudomonas*; leaf-rusts, *Uromyces*; powdery mildew, *Erysiphe polygoni*.

The alphabetic treatment in this volume lists the most serious pests and pathogens under each pulse, as was done also by Summerfield & Roberts (1985). For a general review, see Smartt (1976). Detailed basic and adapted research on diseases and pests is being carried out in regional cooperation (e.g. ICRISAT–CRIFC).

Besides improving soil conditions, rotation of various crops with pulses reduces the danger of pests and diseases. Groundnuts may be rotated with other legume species and should be rotated with species as unrelated as possible. Several pathogens attack both *Vigna* and *Phaseolus*. Also nematodes, the damage from which on pulses has not been widely ascertained for South-East Asia, can be suppressed by rotation.

Cultivated pulses are selected for non-shattering of pods. Wild species shatter their seeds, some less advanced cultivars of *Vigna* and *Phaseolus* species and *Glycine* have the problem of early seed losses too. Frequent wetting and drying during ripening worsens the effects of shattering, and harvests should be frequently repeated to reduce losses.

Harvest is either manual, sometimes by frequent picking, if used as vegetable or, to a small degree, by combine-harvesting. Mechanical harvesting requires cultivation of special cultivars, which mature simultaneously and in a restricted time, especially determinate bush-type cultivars. Small plants are pulled up. Groundnuts may be uprooted with ploughs. Major areas growing soya are fully mechanized but, in the wet tropics, the pulled plants (e.g. mung beans, groundnuts) should be dried before harvesting. Tuberous legumes are generally dug up by hand.

Well winnowed and cleaned seeds must be dried to a moisture content less than 10 % for optimum storage. Legume seeds keep well because of the hard testa, but insect infestation, high temperature and moisture cause deterioration of stored pulses. Bruchids and other seed beetles can be devastating in storage and only hygienic measures may counteract their efforts. Dhal (split peas) keeps better because developing bruchid larvae are provided with less space and do not develop fast in single cotyledons. Various disinfection methods include village remedies such as adding neem leaves (*Azadirachta indica* A. Juss.), chilli powder (but certain storage beetles thrive on chillies), and activated clay. Dusting with insecticides prevents deterioration of sowing seed by bruchids. For food grains, safety limits for health eliminate the use of most insecticides as preservative. Heat treatments provide protection: 57 °C for 3–4 hours has been recommended.

1.6 Breeding and genetic resources

Breeding aims are summarized here. A host of special goals is included but yield is the main concern in breeding. Potential biological yields are much higher than those achieved on farms, even if the ceilings are lower than those of cereals. Potential yields of pulses may be at least four times the national averages. 'Breaking the yield barrier' involves a combination of genetically improved genotypes and agronomic measures, timely application of inputs and labour, and capital investment. Selection for photosynthetically longer active vegetative parts is feasible. Yield is intricately combined with yield stability; so cultivars should be adapted to locality and season. Wide adaptability is an advantage but opens large risks from pests and diseases. So resistance to pests and diseases rank high among aims in breeding. Tolerance to environmental stress used to be a goal for breeding but that idea has usually been replaced by breeding locally adapted cultivars, for instance for (even more) arid or saline conditions. Breeding for poor farming conditions has made no breakthroughs. Breeding for quality factors, such as protein, oil, amino acids or palatability, has been relegated to lower priority, apart from oils from soya beans and groundnuts. The screening for protein content is rather complicated, as influences of locality and season often override genetic effects. A difference of, for instance, 5 %, 18–23 %

between seasons, proves unexceptional, and even larger differences are found. Quality of new cultivars has to be monitored, to keep protein and oil contents acceptable. Consumer preference provides a useful array of visual seed morphology, taste and technological adaptation.

ICRISAT (1987) summarized research priorities for groundnuts, chickpeas and pigeon peas for the various countries. For instance, disease resistance in groundnut should be directed to late leaf-spot in Burma, leaf-spots and rust in China, bacterial wilt in Indonesia, peanut stripe virus in the Philippines and Indonesia, and rust, leaf-spots and *Aspergillus flavus* in Thailand. Free exchange of germplasm and breeding material is imperative, and occurs to various extent through the offices of the national quarantine systems.

IBPGR (1980; 1984; 1985; 1987) lists base and working collections of germplasm for the various crops. Exchange with countries or international institutions outside the South-East Asian Region is frequent. For the major pulses, most diversity has been collected but is not yet always well documented and maintained. Local gaps are many but concerted efforts are under way to counteract the effects of genetic erosion. The collections of minor pulses such as *Lablab*, the lesser *Vigna* spp., *Lathyrus* and *Macrotyloma* are less complete.

Pulse collections are available in several international, regional and national institutes (Tables 5 and 6). Within or near the region, prominent institutions

Table 5. Some major institutions holding pulse germplasm.

Institutions	Crops
International	
CIAT, Cali, Colombia	*Phaseolus* beans, fodder legumes
ICARDA, Aleppo, Syria	Faba beans, fodder legumes
ICRISAT, Patancheru, India	chickpeas, pigeon peas, groundnuts
IITA, Ibadan, Nigeria	cowpeas, *Vigna* spp.
Regional	
AVRDC, Shanhua, Tainan, Taiwan	soya beans, *Vigna* spp., esp. mung beans
INTSOY, Univ. Illinois, Urbana-Champaign, United States	soya beans
National	
BORIF, Bogor, Indonesia	soya beans, groundnuts, mung beans, pigeon peas
CAAS, Oil Bearing Crops Inst., Wuhan, Hubei, China	soya beans
CSIRO, Canberra, Australia	cowpeas, soya beans, vetches, *Vigna* spp., winged beans
Liaoning Agric. Acad., Harbin, China	soya beans
MARIF, Malang, Indonesia	soya beans, groundnuts, mung beans, pigeon peas
NBPGR, New Delhi, India	peas, winged beans, *Vigna* spp.
NPGRL, Univ. Philippines, Los Baños, Philippines	*Vigna* spp., groundnuts, pigeon peas, winged beans
Shadong Agric. Inst., Jinan, Shadong, China	soya beans
Univ. Kebangsaan, Kuala Lumpur, Malaysia	winged beans
USDA Southern Regional Plant Introd. Station, Experiment, Georgia 3042, United States	groundnuts, *Vigna* spp.

Table 6. Number of collections of cultivated food legumes (small is 1–50, medium 51–250 and large > 250 accessions) and their wild relatives in the world (IPBGR, 1980)[1].

	Cultivated			Wild	Total
	small	medium	large		
Arachis hypogaea	.	1	7	3	11
Cajanus cajan	1	.	2	1	4
Canavalia spp.	.	2	.	.	2
Cicer arietinum	5	1	7	4	17
Glycine max	.	.	13	7	20
Lablab purpureus	1	1	.	.	2
Lathyrus spp.	2	2	1	.	5
Lens culinaris	4	3	4	3	14
Lupinus spp.	1	5	4	.	10
Phaseolus acutifolius	1	.	.	.	1
Phaseolus coccineus	.	1	1	.	2
Phaseolus lunatus	.	.	1	.	1
Phaseolus vulgaris	.	1	14	.	15
Phaseolus spp.	2	4	13	.	19
Pisum sativum	3	5	11	.	19
Psophocarpus tetragonolobus	2	1	4	.	7
Trigonella spp.	2	.	.	.	2
Vicia faba	3	3	6	.	12
Vicia spp.	1	.	5	.	6
Vigna angularis	.	1	1	.	2
Vigna mungo	.	1	1	.	2
Vigna radiata	.	.	5	.	5
Vigna umbellata	.	1	1	.	2
Vigna unguiculata	1	2	5	.	8
Vigna spp.	1	2	1	.	4

1. Some important collections were excluded if number of accessions was not specified. The 1980 issue is being updated.

· = not registered

are the Chinese Academy of Agricultural Sciences (CAAS), Beijing, with its regional stations; the Asian Vegetable Research and Development Center (AVRDC), Taiwan; the Bogor Research Institute for Food Crops (BORIF) in Bogor and the Malang Research Institute for Food Crops (MARIF) in Malang, Indonesia; the National Plant Genetic Resources Laboratory (NPGRL) and the University of the Philippines, Los Baños, the Philippines; the National Bureau for Plant Genetic Resources (NBPGR), New Delhi; the International Crops Research Institute for the Semiarid Tropics (ICRISAT), Patancheru, India; and the Universiti Kebangsaan Malaysia, Kuala Lumpur. Each institute is specialized in one crop or several crops. Base collections are kept at −20 °C but those conditions are not yet operational everywhere. Active (working) collections are preferably stored at +4 °C and such storage usually exists. At −20 °C and +4 °C, seeds remain viable at least 100 and 25 years, respectively.

1.7 Prospects

Emphasis in research is now given to improvement of cereal genotypes and their place in agriculture. However pulses are attracting attention too. To diversify and upgrade nutritional value of the diets of the people in South-East Asia, pulses must play a large role; their contribution to the supply of protein and fats as well as the balance pulses provide to the amino acid profile of cereal protein make them valuable. With proper inputs, the future of pulses will be bright. Many national efforts are indeed being made and international agencies are encouraging research initiatives. Economic incentives are favourable, as prices of pulses and oilseeds are favourable.

The main Grain Legumes Network, an initiative of ICRISAT, seeks to promote research on groundnuts, chickpeas and pigeon peas, through exchange of information and breeding material (ICRISAT, 1984; 1987).

Intensive cropping, multiple-cropping seasons with well adapted short-duration cultivars, better management, expansion to less suitable environments all hold promise for increase and diversification of production of pulses. Pulses and related forage legumes are perhaps the best choice to improve soil in agriculture, expecially where legume–soil–*Rhizobium* interaction can be optimized. Yield potentials are far above 2 t/ha (1 t/ha = 100 g/m^2). Sustainable yields need further work on breeding and inputs from agronomy.

2 Alphabetical treatment of species

Arachis hypogaea L.

Species Plantarum 2: 741 (1753).
LEGUMINOSAE
$2n = 4x = 40$

Vernacular names Groundnut (En). Peanut (Am). Arachide (Fr). Indonesia: kacang tanah, kacang Jawa, kacang Manila. Malaysia: kacang China, kacang Jawa, kacang goreng. Papua New Guinea: galip bilong giraun, kasang. The Philippines: mani, batung-China. Cambodia: sânndaèk dèi. Laos: thwàx din, thwàx ho. Thailand: thua lisong, thua din. Vietnam: dâu phong, lac.

Origin and geographic distribution Groundnuts originated in the area southern Bolivia – north-western Argentina of South America. The Portuguese apparently took them from Brazil to West Africa and then to south-western India in the 16th Century. Africa is now regarded as a secondary centre of diversity. At the same time too, the Spaniards introduced them from Mexico to the western Pacific whence they spread to China, Indonesia and to Madagascar. The Dutch also probably took them from Brazil to Indonesia by the middle of the 17th Century. They were probably introduced to the United States on slave ships from Africa, though they may have been introduced directly from the Caribbean Islands. Groundnuts are now grown in most tropical, subtropical and temperate countries between 40°N and 40°S latitude, especially in Africa, Asia, North and South America. In Asia, the groundnut is a major crop in India, China, Indonesia, Burma, Thailand and Vietnam.

Uses Most of the world crop is crushed for oil that is used mainly for cooking. The presscake from oil extraction is a feed rich in protein but is also used to produce groundnut flour, which is used in many human foods. Most of the Burmese crop, about 20 % of the Indonesian crop and 30 % of the Thai crop is crushed for oil. However most of the crop in South-East Asia is used as such for human consumption. The seeds or kernels are eaten raw, boiled or roasted, made into confectionery and snack foods, and are used in soups or made into sauces to use on meat and rice dishes. The vegetative residues from the crop are excellent forage.

Production and international trade Groundnuts are a major world oilseed and food legume. The following data are based on FAO statistics for the period 1981–1985. In that period, the world area under cultivation averaged 20 million ha and in-shell production averaged 19 million t. From trade figures over that period, the world price was estimated to be US$ 860 per tonne on a fresh shelled basis.
By country, Indonesia produced 792 000 t from 500 000 ha, Burma produced 600 000 t from 585 000 ha, Thailand 149 000 t from 127 000 ha, Vietnam 92 000 t from 136 000 ha, the Philippines 48 000 t from 55 000 ha, Malaysia 21 000 t from 6000 ha, Laos 10 000 t from 12 000 ha and Cambodia 9000 t from 8000 ha. Production in Papua New Guinea was less than 1000 t. The Indonesian crop is produced entirely by smallholders.
Most production in each country is consumed locally. Only Vietnam (21 000 t) and Thailand (5800 t) were significant exporters of groundnut; Indonesia imported 30 000 t and Malaysia 9400 t during the period 1981–1985.

Properties Per 100 g edible portion groundnuts contain roughly: water 5.4 g, protein 30.4 g, fat 47.7 g, carbohydrates 11.7 g, fibre 2.5 g and ash 23 g. The energy content averages 2457 kJ/100 g. Groundnuts are good sources of vitamins B and E. Seed weight varies between 25 and 115 g/100 seeds. Groundnut kernels are noted for their high contents of oil and of protein. Content of crude protein in whole dry kernels ranges between 22 and 30 % and varies with type, cultivar, locality, year and physiological maturity of the seed. Aspartic acid, glutamic acid and arginine account for about 45 % of total amino acids. Lysine, methionine and threonine tend to be deficient.
Content of oil in whole dry kernels ranges between 44 and 56 % with an average of 50 %. Cultivars of Virginia type tend to have a lower content of oil than Spanish types. Groundnut oil is an unsaturated liquid susceptible to oxidation and hence rancidity. Oleic and linoleic acids are unsaturated acids and together account for about 80 % of the fatty acid. Palmitic acid accounts for a further 10 % and the other common fatty acids range in concentration from 0.01 to 4 %. The higher the ratio of oleic to linoleic acids in groundnut oil, the more stable is the oil; that is, it is less prone to oxidative rancidity. The ratio in mature kernels can range from less than 1.0 to greater than 3.0, more than 1.3 generally being considered satisfactory by processors. Cultivars of Spanish type tend to have lower ratios than those of Virginia type.

Description A monoecious, prostrate to erect annual herb, usually 15–70 cm high. Root system consisting of a well developed tap-root with many lateral roots, able to penetrate to depths in excess of 2 m; root hairs absent, nitrogen-fixing nodules present. The main stem or central axis develops from the epicotyl and bears a cotyledon at each of

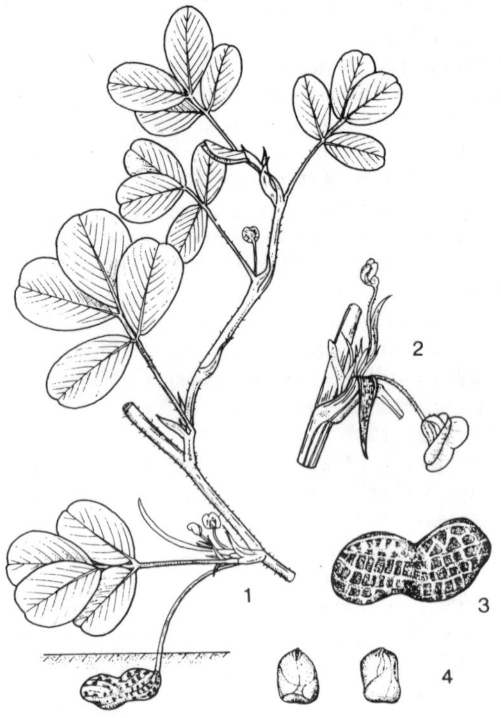

Arachis hypogaea L. – 1, branch with flowers and fruit; 2, inflorescence; 3, fruit; 4, seeds.

the first two nodes. Branching dimorphic, with vegetative branches and contracted reproductive branches. All vegetative branches with scale leaves, called cataphylls, subtending the first two nodes of the branch. Secondary and tertiary vegetative branches can develop from the primary vegetative branches. Leaves on the central stem spirally arranged in a 2/5 phyllotaxy, on the primary vegetative branches arrangement is distichous; leaves are 4-foliate with two opposite pairs of obovate leaflets of 3–7 cm × 2–3 cm; petiole 3–7 cm long; pulvini at the base of the petiole and at the base of each leaflet cause characteristic night movements where the petiole folds downwards and the leaflets fold upwards until they touch.

The contracted reproductive branches or inflorescences are formed singly at both cataphyllar and ordinary leaf axils on vegetative branches and, in some forms, at higher nodes on the central stem; each inflorescence bears 2–5 flowers; short branches develop in the axils of simple bracts that occur along the central axis of the inflorescence and terminate after production of one leaf, the bifid bract, in the axil of which the flower is borne; an inflorescence never occurs in the same leaf axil as a vegetative branch. Flowers sessile, consisting of a 4–6 cm long tubular hypanthium (fused lower parts of calyx, corolla and staminal tube), the top of which bears expanded lobes of the 5 sepals and petals (pale yellow through to orange-red), and 10 short filaments with anthers; superior ovary of single sessile carpel with 2–6 ovules, situated at the base of the hypanthium tube; style free within the tube, ending in a club-shaped stigma among the anthers.

After fertilization, a stalk-like structure called 'peg' or gynophore elongates with an intercalary meristem at the base of the ovary. The peg grows down in a curve towards the soil, carrying the ovary at its tip, which becomes hardened into a protective cap as the peg enters the soil. The length of the peg depends on the initial distance of the flower from the soil but, if more than 15 cm, it usually fails to reach the ground and the tip dies. When the peg has penetrated the soil to a depth of 2–7 cm, it turns horizontal and fruit development begins as the tip swells rapidly. Mature fruits (pods) cylindrical, 1–8 cm × 0.5–2 cm, containing 1–6 seeds. The pod surface or pericarp may be constricted to varying degrees between the seeds and is reticulate with veins in the hardened mesocarp. The mature seeds or kernels are cylindrical to ovoid, 1–2 cm × 0.5–1 cm. Each seed is enclosed in a thin papery testa ranging in colour from white to pink, red, purple, tan and shades of brown. Each seed has two large cotyledons, an epicotyl with leaf and bud primordia, a hypocotyl and the primary root. Upon epigeal germination, the primary root elongates rapidly, reaching 10–12 cm before lateral roots appear. As growth proceeds, the outer layer of the primary root of a seedling is sloughed off so that root hairs do not form.

Growth and development After harvest, seed of cultivars of Virginia type have a dormant period of 1–6 months, depending on temperature and storage conditions. Heat treatment or ethylene can break dormancy. Cultivars of Spanish and Valencia types do not possess seed dormancy. After sowing, seedlings emerge in 5–10 days, depending on sowing depth and soil temperature. A period of vegetative development ensues for 20–35 days, before the first flowers appear. Groundnuts can flower over a long period (20–60 days), depending on plant water status, irradiance, temperature and photoperiod. That flowering period is considerably shorter in Spanish and Valencia cultivars than in Virginia cultivars. Daily production of flowers typically increases to a peak and then declines. On each inflorescence, only one flower opens on one day and a period of one to several days elapses before another flower opens on the same inflorescence. The flower bud elongates rapidly during the night before anthesis. Anthesis and self-pollination within the closed flower occurs around sunrise the following day. Normally 60–80 days are required for pod development from flowering to maturation in Virginia types and slightly less than that (50–60 days) in Spanish types. As groundnuts are botanically indeterminate, vegetative development can continue during flowering and pod filling, though vegetative development typically declines or ceases during pod filling. In the warm tropics, short-season cultivars of the Spanish–Valencia group mature in 85–100 days whereas longer-season cultivars of the Virginia group require 110–130 days. In cooler subtropical environments or at a height, corresponding periods are 115–130 days for Spanish types and 140–160 days for Virginia types.

Other botanical information There is considerable variation in *Arachis hypogaea* and two subspecies have been proposed: ssp. *hypogaea* and ssp. *fastigiata* Waldron. Breeding programmes have crossed cultivars from the two subspecies and some subspecific traits such as branching pattern are no longer distinct in cultivars developed from such crosses.

- Ssp. *hypogaea*, the Virginia group, is characterized by a central axis that never bears inflorescences and that has lateral branches where two vegetative branches alternate regularly with two inflorescences or reproductive branches; cultivars are late maturing; plants are dark green with a prostrate to spreading bunchy habit; pods are typically two-seeded and seeds show marked dormancy.
- Ssp. *fastigiata*, the Spanish–Valencia group, is characterized by inflorescences on the central axis and without a regular pattern in the sequence of reproductive and vegetative branches; cultivars mature earlier; plants are lighter green (Spanish) with erect growth habit; pods are concentrated around the central axis, contain 2–6 seeds without dormancy. Spanish types usually have two-seeded pods; Valencia types typically have pods with 3–6 seeds, thicker stems and considerably fewer secondary and tertiary branches than Spanish types.

Ecology Groundnuts are grown between latitudes 40 °N and 40 °S in the warm tropics and subtropics, and in temperate humid regions with sufficiently long warm summers. Optimum mean daily temperature to grow is 30 °C and growth ceases at 15 °C. The phenology of groundnuts is determined primarily by temperature, with cool temperatures delaying flowering. In controlled environments, photoperiod has been shown to influence the proportion of flowers producing pods and distribution of assimilates between vegetative and reproductive structures (harvest index) in some cultivars. Long photoperiods (greater than 14 hours) generally increase vegetative growth and short photoperiods (less than 10 hours) increase reproduction growth. However the significance of these effects in crop communities in the tropics and subtropics has yet to be established.

Between 500 and 600 mm of water reasonably well distributed through the growing season allows satisfactory production. Nevertheless, groundnuts are a drought-tolerant species and can withstand severe lack of water, though yield is generally reduced.

Because pods develop underground and must be recovered at harvest, friable well drained soils are preferred, though plants grow and develop adequately on heavier clay soils. For optimum growth, pH should be in the range 5.5–6.5, though Spanish types tolerate more acid conditions (to pH 4.5) and some cultivars grow well in alkaline soils to pH 8.5.

Propagation Commercial crops are grown from seed. Ideally the seed-bed should be deep and friable with an even particle size, and should be weed-free. In some countries, seed is routinely dressed with a protective fungicide and sown with mechanical planters. Cloddy and uneven seed-beds, and hand-planting can result in uneven emergence and heavy seed-bed losses of plants. In South-East Asia, groundnuts are grown mainly by smallholders either as a rain-fed crop during the wet season or in lowland areas after rice as a second or third crop either with supplementary irrigation or on residual moisture. They are grown as a sole crop and also intercropped with maize, soya beans and cassava. In some areas, they are grown under perennial tree crops such as coconut, oil palm or rubber.

Recommended plant densities are near 200 000–250 000 plants/ha for the typically short-season Spanish cultivars. That contrasts with tropical areas of Australia where densities of 100 000–125 000 plants/ha are recommended. In most countries, cultivation is in rows with plant spacings ranging from 40 cm × 20 cm to 30 cm × 20 cm.

Husbandry To achieve maximum economic yields, most competing weeds must be eliminated.

In South-East Asia, that is often attempted by hand-weeding. However a range of pre-emergence and post-emergence herbicides are available to control grasses and many broad-leaf weeds.

The only peculiar nutrient requirement is for Ca in the podding zone. Calcium is absorbed directly by the pods, if soil moisture is adequate. A shortage of Ca in that zone will result in empty pods, particularly in cultivars of Virginia type. The crop's needs for nitrogen should be satisfied with symbiotic fixation by strains of *Rhizobium* of the cowpea group. So nitrogen fertilizers are not generally required. In some areas of acid soils, lime is applied to raise the pH.

Moisture stress during flowering or pod filling reduces yield so that irrigation during those periods to minimize or eliminate the stress increases production and seed quality.

Diseases and pests The most serious fungal diseases of foliage are leaf-spots and rust. Early leaf-spot (*Cercospora arachidicola*, *Mycosphaerella arachidis*), late leaf-spot (*Cercosporidium personatum*, *M. berkeleyii*) and rust (*Puccinia arachidis*) can cause significant losses, particularly during the wet season. The diseases can be controlled with fungicides. Resistant cultivars are not generally available at that stage. The soil-borne fungus *Aspergillus flavus* and related species are widespread in the region and infected groundnuts can be contaminated with the carcinogen aflatoxin. Data from Indonesia and Thailand suggest that *A. flavus* as a mould contaminant and toxin producer is much less serious during growth of the crop than during subsequent storage of kernels. Minimizing moisture stress during growth can reduce invasion and toxin production by *A. flavus*. Bacterial wilt (*Pseudomonas solanacearum*) is serious in Indonesia, Malaysia and China. It can be controlled by use of resistant cultivars.

Peanut stripe virus (PStV) is a serious disease in Indonesia, Thailand and the Philippines. It can cause serious losses, if a crop is infected early. The virus is transmitted by aphids (*Aphis craccivora*) and seed (up to 30 % seed transmission). No control measures are currently available. Peanut mottle virus (PMV) is widespread but is considered less serious because of the lower incidence of seed transmission and less severe effects on yield. Peanut yellow spot virus (PYSV) is widespread in Thailand.

Nematodes, particularly species of the genera *Meloidogyne* and *Pratylenchus*, are widespread but their seriousness for yield is unknown. Control measures include nematicides and crop rotation.

A wide range of insects can attack groundnuts and can sometimes cause severe losses. Leaf miners (*Aproaerema modicella*) have caused substantial damage to foliage in Indonesia, Thailand and the Philippines. Other insects, often present in large numbers, include leafhoppers (*Empoasca* spp.), leaf-eating caterpillars (*Heliothis* spp.) and thrips (*Frankliniella schultzei*, *Scirtothrips dorsalis*). Control by insecticides is possible. Aphids (*Aphis craccivora*) as virus vectors are a serious pest, particularly in the dry season.

Harvesting Spanish cultivars are harvested 85–100 days after sowing and Virginia cultivars 110–130 days after sowing in the warm tropics. Severe disease of foliage sometimes results in harvesting before seeds are fully mature. Much is harvested by hand in South-East Asia. Plants are pulled from the ground and pods removed from the bushes. Pods are then sun-dried to about 10 % moisture. Where machines are available, the tap-roots are cut and plants are lifted mechanically. After 2–10 days of sun-drying, pods are mechanically threshed from the bushes.

Yield Average yield of pods in South-East Asia is 1 t/ha, though considerable variation in yields between countries is reported: 0.7–0.9 t/ha for Vietnam, Laos, the Philippines and Papua New Guinea; 1.0–1.2 t/ha for Thailand, Burma and Cambodia; 1.6 t/ha for Indonesia; 3.6 t/ha for Malaysia. Average yield for the world is 0.95 /ha, though in the United States average yield is 2.9 t/ha.

Handling after harvest To minimize the development of *A. flavus* and subsequent toxin production, groundnuts should be dried to less than 14 % moisture. Interruption and retardation of sun drying by showers or overcast humid weather, or moisture uptake during storage can result in growth of *A. flavus* and contamination with aflatoxin. Seeds can be protected from mechanical damage by storage and transport in the shell. In many areas, pods are sold directly to local consumers to provide cash for farmers. Local hand-shelling is common too. In Thailand, machines are used to shell, extract oil and grind. Processing factories produce dried or boiled groundnuts or groundnut products.

Genetic resources The International Crops Research Institute for the Semiarid Tropics (ICRISAT) at Hyderabad in India maintains the largest world germplasm collection of over 12 000 accessions. ICRISAT has been designated as a principal repository for *Arachis* germplasm by the International Board for Plant Genetic Resources. The

United States Department of Agriculture maintains an extensive germplasm collection too. The Bogor Research Institute for Food Crops (BORIF) in Indonesia maintains a collection of local Indonesian germplasm and some introductions.

Breeding Development of higher-yielding cultivars adapted to environments and to production systems used in the region is the major objective of national and international (ICRISAT) breeding programmes. Earliness (less than 80 days) and drought tolerance are objectives in rice-based farming systems in Indonesia and Thailand. Other objectives include resistance to rust, leaf-spots, bacterial wilt (Indonesia) and *A. flavus* (Thailand), seed dormancy (Indonesia and Thailand) and tolerance of acid soils (the Philippines and Indonesia). Currently the world germplasm collection is being screened in Indonesia to find sources of resistance to peanut stripe virus.
A. hypogaea is an allotetraploid. Several diploid wild species, including *A. cardenasii* Krap. et Greg. (nom. ined.) and *A. chacoense* Krap. et Greg. (nom. ined.), are being used as sources of disease resistance at ICRISAT and in the United States.

Prospects Groundnut seems to have bright prospects. As a short-season annual tropical legume, it can contribute to the nitrogen economy of the associated farming systems and provide a valuable source of protein for man. Because groundnut is a valuable commodity, it can provide cash for smallholders too.
Farmers generally perceive groundnuts and other food legumes to have a lower yielding and to give a lower return than cereals with low inputs and marginal cultural conditions. However the significance of the crop in farming systems and as a source of additional income is well recognized. The crop is well established in ecological niches of various cropping systems and there are good prospects of expanding the role of groundnuts in those systems and of inclusing groundnuts in new cropping systems. Considerable efforts are being made to increase groundnut production in most South-East Asian countries.

Literature 1 Gregory, W.C., Krapovickas, A. & Gregory, M.P., 1980. Structures, variation, evolution and classification in Arachis. In: Summerfield, R.J. & Bunting, A.H. (Editors): Advances in legume science. Royal Botanic Gardens, Kew. pp. 469–481. **2** ICRISAT, 1980. Proceedings of the International Workshop on Groundnuts, 13–17 October 1980. Patancheru, A.P., India. 325 pp. **3** ICRISAT, 1987. Groundnut rust disease. Proceedings of a discussion group meeting, 24–28 September 1984. Patancheru, A.P., India. **4** Pattee, H.E. & Young, C.T. (Editors), 1982. Peanut science and technology: speciation and cytogenetics in Arachis; genetics of Arachis hypogaea L.; breeding of the cultivated peanut; cultural practices; management of preharvest insects; peanut plant diseases; harvesting, curing and energy utilization; aflatoxins and other mycotoxins in peanuts; insect control in postharvest peanuts; composition, nutrition and flavor of peanuts. American Peanut Research and Education Society, Texas. 825 pp. **5** Porter, D.M., Smith, D.H. & Rodriguez-Kabana, R. (Editors), 1984. Compendium of peanut diseases. American Phytopathological Society. 73 pp. **6** Purseglove, J.W., 1968. Tropical crops. Dicotyledons. 1. Longmans, Green & Co. Ltd, London. pp. 225–236. **7** Smartt, J., 1985. Evolution of grain legumes. 5. The oilseeds. Experimental Agriculture 21:305–320.

R. Shorter & A. Patanothai

Cajanus cajan (L.) Millsp.

Field Columbian Museum, Botanical Series 2(1): 53 (1900).
LEGUMINOSAE
$2n = 22, 44, 46$

Synonyms *Cytisus cajan* L. (1753), *Cajanus indicus* Spreng. (1826).

Vernacular names Pigeon pea (En). Pois d'Angole, ambrévade (Fr). Indonesia: kacang Bali, kacang gude, kacang kayu. Malaysia: kacang, kacang dal, kacang hiris. The Philippines: tabios, kardis, kidis. Cambodia: sândaèk dai, sândaèk kroëb sâ, sândaèk klöng. Laos: thwàx h'ê. Thailand: thua rae, thua maetaai, ma hae.
Vietnam: cay dau chieu, dau sang, dau thong.

Origin and geographic distribution Pigeon pea originated in India and spread to South-East Asia in the early centuries of our era. It reached Africa 2000 BC or earlier, and found its way to the Americas with the conquests and slave trade, probably through both the Atlantic and the Pacific. It is now grown all over the tropics but especially in the Indian Subcontinent and East Africa.

Uses In contrast to the Indian Subcontinent, where pigeon pea is mainly used as a pulse (dhal = split pea), the use of fresh seeds and even pods as a vegetable in sayors (spicy soups) and other side-dishes preponderates in South-East Asia. Ripe seeds are eaten roasted too. Pigeon pea may replace soya beans to make tempeh and tahu (fermented products). Pigeon pea is useful as tall hedges on dry soil and on the bunds of paddy fields. The branches and stems can be used for baskets and fuel. It is often grown as a shade crop, cover crop or windbreak, or even as support for vanilla. After establishment, pigeon pea improves the soil by its extensive root system, nitrogen fixation by *Rhizobium* and the mulch provided by the fallen leaves. It may serve as host for silkworms (Madagascar) or the lac insect (northern Bengal, Thailand). Traditional uses as medicine are many, e.g. young leaves are applied to sores, herpes and itches in Java.

Production and international trade Pigeon pea covers 3 million ha, of which 85 % are in India, and produces 2 million t/year. Major areas of production are eastern Africa and the Caribbean. In almost 50 countries, the crop is of local importance. In South-East Asia, it is occasionally employed as a hedge or garden crop. No reliable statistics exist but the plant can yield well in drier areas.

Properties Per 100 g edible portion dry seeds contain: water 7–10.3 g, protein 14–30 g, fat 1–9 g, carbohydrates 36–65.8 g, fibre 5–9.4 g, ash 3.8 g. The energy content averages 1450 kJ/100 g. Cooking time of dhal is 24–68 minutes, smaller grains are ready sooner. Fresh seeds contribute vitamins, especially provitamin A and vitamin B complex. Seed weight varies between 4 and 26 g/100 seeds.

Description A glandular-pubescent, short-lived perennial (1–5 years) shrub, usually grown as an annual, 0.5–4 m high, with thin roots up to 2 m deep. Stems up to 15 cm in diameter. Branches many, slender. Leaves alternate, trifoliolate, glandular punctate; leaflets elliptical, 3–13.7 cm × 1.3–5.7 cm. Flowers in pseudoracemes, sometimes concentrated and synchronous (determinate), usually scattered and flowering over a long period (indeterminate), papilionaceous, corolla yellow or cream, standard dorsally red, orange or purple. Fruit a straight or sickle-shaped pod with (2–)4–9 globose to ellipsoid or squarish seeds. Seeds white, cream, brown, purplish to almost black, plain or mottled; strophiole usually virtually absent. Seedlings with hypogeal germination; first leaves simple.

Growth and development Emergence is complete 2–3 weeks after sowing. Vegetative development starts slowly. After 2–3 months, growth accelerates. Flowering (of half the plants) starts 56–210 days after sowing. Maturity ranges from 95 to 256 days in normal conditions with rainy season and long days. With short days, growth in length

Cajanus cajan (L.) Millsp. – 1, branch with inflorescence; 2, fruits; 3, seed.

is less and flowering accelerated. In Indonesia, flowering and fruiting may continue throughout the year.

Other botanical information Ten maturity groups may be distinguished under Indian conditions, usually combined into four categories: extra early, early, medium and late-maturing cultivars 120, 145, 185 and more than 200 days after sowing, respectively. Continuous variation in present-day world collections shows that formal varieties, var. *flavus* (DC.) Purseglove and var. *bicolor* (DC.) Purseglove, described by De Candolle as species, can no longer be maintained.

Ecology Flowering is triggered by short days and plants grow vegetatively with long days, as in the rainy season of India. There are few truly day-neutral forms. Optimum temperatures range from 18 to 38 °C, frost is not tolerated. Above 29 °C, soil moisture and fertility need to be adequate. Rainfall optimum is 600–1000 mm/year, waterlogging is harmful. Pigeon pea is rarely found above altitude 2000 m. Drained soils of reasonable water-holding capacity and with pH 5–7 or more are favourable. The plant tolerates an electrical conductivity (salinity) from 0.6 to 1.2 S/m.

Propagation Propagation is by seed. Stem cuttings rarely succeed.

Husbandry Seeds should be sown in rows with spacing 30–50 cm × 75–150 cm. In intercropping, the crop performs well with 2 rows of cereals (e.g. sorghum, millets), cotton or groundnut. After harvest of the intercrop, long-duration pigeon pea continues to grow and protects the soil. As a field crop, pigeon pea may be typified as rather primitive, particularly the tall genotypes being cumbersome in cultivation. Weeds must be controlled to alleviate slow initial growth. Wind may bend the plants but staking is not practised. Irrigation as a life-saver can be economic. In intensive cropping of short-duration cultivars, irrigation may be required. Mechanization is only possible for short cultivars. Response to fertilizers is rarely economic; a phosphate dressing is generally recommended at 20–100 kg/ha.

Diseases and pests Because of its long flowering period, pests such as *Heliothis* borers and *Agromyza* fruitflies may be compensated for by renewed flushes. Chemical control is cumbersome and expensive in tall indeterminate forms. Crop rotation is advisable against diseases such as *Fusarium* wilt.

Harvesting The crop is usually cut near the ground when most pods are mature. Many leaves are then still green. Green pods are picked over a long period in home gardens or hedge crops. Ripe pods can be harvested with combine-harvesters but only for cultivars maturing uniformly with pods at a uniform level above the ground.

Yield In India, yield averages 716 kg/ha. In marginal areas, yield is 700 kg/ha in sole cropping, but with optimum conditions, yields of more than 5000 kg/ha are possible. In intercropping with maize in Indonesia, low yields of 175 kg/ha were obtained but, in the eastern part of Indonesia, sole cropping may produce 3000–4000 kg/ha.

Handling after harvest Entire air-dried plants are threshed, usually by hand or with cattle, and seed is cleaned. Clean bins prevent insect attack, which can be considerable. Storage as split peas reduces bruchid attack. Processing includes dhal making, either wet (after sprinkling heaps of seed) or dry, by milling. In the West Indies, canning and freezing of fresh pigeon peas is a million-dollar export business, for instance to the United States.

Genetic resources The world germplasm collection has covered India and several African countries, and some Caribbean islands for a second time. More than 11 000 samples are available in the ICRISAT collection near Hyderabad, India, and various breeders and institutes have parts of that collection. Attempts are continuing to cover all areas of occurrence.

Breeding High yield and consumer and miller preference are prime criteria. Stability of yield may be obtained by selecting for photoperiod insensitivity, disease and pest resistance, suitability for intercropping and for multiple harvests. For most of those characteristics, improved genotypes are now available. Resistance is available in wild relatives and there are promising pest-resistant and disease-resistant genotypes. Short-duration Indian cultivars include 'Prabhat', 'Pusa Ageti', 'Sharda', 'T21', 'UPAS-120'; good medium-duration cultivars are 'C 11', 'BDN-1', and several ICP lines. Hybrid cultivars exist too, including 'ICPH 2, 5, 6, 8'. Several wild relatives, e.g. *Cajanus albicans* (W. & A.) Maesen, *C. sericeus* (Benth. ex Bak.) Maesen, *C. scarabaeoides* (L.) Thouars, cross with pigeon pea, the closest one is *C. cajanifolius* (Haines) Maesen (= *Atylosia cajanifolia* Haines). Hybrids have contributed male sterility, but the transfer of insect resistance (from *C. scarabaeoides*), high content of protein (several species), improved drought resistance (*C. acutifolius* (Benth.) Maesen) and annuality (*C. platycarpus* (Benth.) Maesen) have not yet materialized. In Malaysia, Singapore, Indonesia, Brunei, the Philippines and Papua New Guinea, the following wild species are found: *Cajanus crassus* (Prain ex King) Maesen, *C. goensis* Dalz., *C. platycarpus* (Benth.) Maesen, *C. reticulatus* (Dryander) F. von Muell. var. *grandifolius* (F. von Muell.) Maesen, *C. scarabaeoides* (L.) Thouars, *C. volubilis* (Blanco) Blanco. Accessions from South-East Asia are still missing from gene banks.

Prospects As a multipurpose crop, pigeon pea is well known but ought to be promoted especially in more semiarid regions of Indonesia (East Java, Sunda Islands) and the Philippines. It fits in smallholders' garden cropping and along hedges and bunds of rice fields.

Literature |1| Dahiya, B.S., 1980. An annotated bibliography of pigeon pea 1900–1977. ICRISAT, Patancheru, A.P., India. 183 pp. |2| ICRISAT, 1984. Grain legumes in Asia. Summary proceedings of the Consultative Group Meeting for Asian Regional Research on Grain Legumes (groundnut, chickpea, pigeonpea). ICRISAT, 11–15 Dec. 1983. ICRISAT, Patancheru, A.P., India. 98 pp. |3| Nene, Y.L. & Kumble, V. (Editors), 1981. Proceedings of the International Workshop on Pigeonpeas. 15–19 Dec. 1980, Patancheru, A.P., India. ICRISAT,

India. Vol. 1 (508 pp.) and Vol. 2 (451 pp). |4| van der Maesen, L.J.G., 1985. Cajanus DC. and Atylosia W. & A. (Leguminosae). A revision of all taxa closely related to the pigeonpea, with notes on other related genera within the subtribe Cajaninae. Agricultural University Wageningen Papers 85-4. 225 pp. |5| Whiteman, P.C., Byth, D.E. & Wallis, E.S., 1985. Pigeonpea. In: Summerfield, R.J. & Roberts, E.H. (Editors): Grain legume crops. Collins, London. pp. 658–698.

L.J.G. van der Maesen

Cicer arietinum L.

Species Plantarum 2: 738 (1753).
LEGUMINOSAE
$2n = 16$, but 14, 24, 32, 33 have been reported too.

Vernacular names Chickpea, Bengal gram, garbanzo bean (En). Pois chiche (Fr). Indonesia: kacang Arab, kacang kuda. Thailand: thua hua chang.

Origin and geographic distribution Chickpea originated in south-eastern Anatolia (Turkey) and reached the Indian Subcontinent before 2000 BC. India, Pakistan, Ethiopia, Turkey and Mexico have the largest areas under chickpea. Around the Mediterranean and in the Middle East, local production is significant. In South-East Asia, chickpeas are occasionally grown in areas with a dry season.

Uses Chickpea is mainly consumed as a dry pulse. Green pods are shelled for the peas and eaten as snack or vegetable. The seed husks are used to feed stock and poultry.

Production and international trade Chickpea is grown on 10 million ha of which 74.5 % in India and 1.7 % in Burma. World production is 5.6 million t/year with India, Pakistan, Mexico, Ethiopia and Turkey as largest producers. In production, chickpea ranks as the third non-oilseed grain legume in the world, after *Phaseolus* beans and peas.

Properties Dry seeds contain per 100 g edible portion: water 7–11 g, protein 12–31 g, fat 4–10 g, carbohydrates 58–68 g, fibre 3–5 g, ash 2–5 g. Energy content averages 1520 kJ/100 g. Kabuli cultivars cook faster and have less dietary fibre than Desi cultivars with coloured seed-coat. Fresh and sprouted seeds contribute vitamin C to the diet. Seed weight varies between 5 and 75 g/100 seeds.

Botany A branched annual with conspicuous acid-producing glandular hairs, usually 20–60 (–100) cm tall with straight or bent stems; roots reaching 1–2 m deep, well nodulated. Leaves alternate, with 11–15 toothed leaflets; leaflets ovate to elliptic, 6–20 mm × 3–14 mm. Inflorescences racemes with 1(–2) flowers; pedicels recurved when fruiting; flowers papilionaceous; corolla white, pink to purplish (fading to blue) or blue. Fruit an inflated rhomboid-ellipsoid pod, 14–29 mm × 8–20 mm with 1–2(–4) seeds. Seed globular to angular obovoid with conspicuous beak, creamy to brown, green or black, surface smooth or wrinkled. Seedlings emerge 7–14 days after sowing, germination is hypogeal, the first two leaves are simple scales. First flowering in less than a month for short-duration cultivars with long days; strict self-pollination. The crop matures in a month; the crop cycle varies from 3 to 6 months. Two or three large cultivar groups can be discerned: the large-seeded cream-coloured wrinkled Kabuli; the small-seeded darker-coloured smooth or wrinkled Desi; some intermediate cultivars.

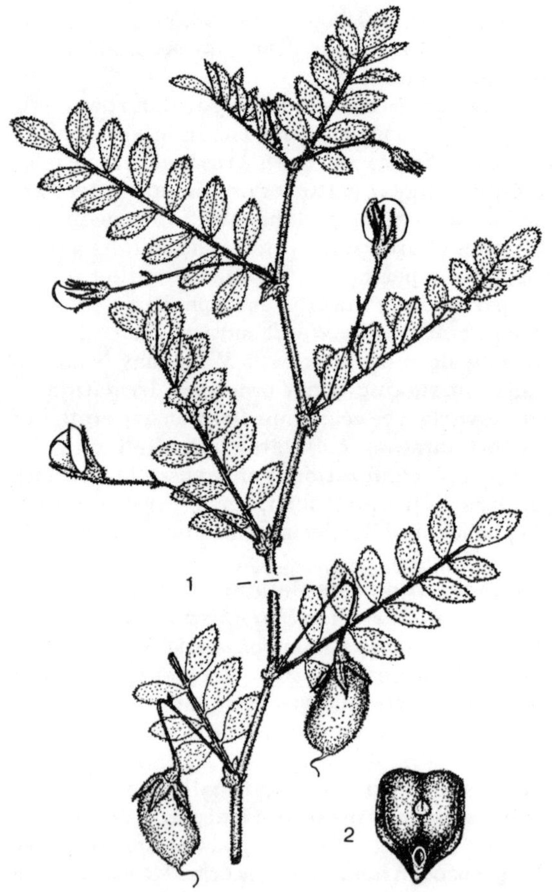

Cicer arietinum L. – 1, branch with flowers and fruits; 2, seed.

Ecology Chickpeas are long-day cool-season plants and grow in semiarid conditions on residual moisture. Drought resistance varies from moderate to considerable. In the Mediterranean, daylengths increase during the season; in India, they decrease. Optimum temperatures range from 15–29 °C. Mild winter or spring rains during the vegetative stage are advantageous. Rainstorms during flowering harm the crop and so monsoon seasons are unsuitable. Soils need to be well drained, pH 5–7 or more, and salinity is hardly tolerated, if at all. Soils vary from sandy to sandy loam and black cotton soils.

Agronomy Propagation is by seed. Rates are 30–50 kg/ha. Optimum plant population is 33 plants/m². Chickpea is mainly a crop for smallholders but large estates in Ethiopia, Spain and Turkey export the crop. Spacing is 30–60 cm between rows and is 10 cm between plants. Intercropping and sole cropping are practised. As a rotation crop, chickpea is ideal. Weed control and sparse irrigation favour chickpeas. Cultivars are rarely responsive to fertilizers but phosphates at 30–50 kg/ha maintain soil fertility; starter nitrogen at 40 kg/ha may be useful. *Heliothis* podborers and *Agromyza* podflies are the main pests; *Ascochyta* blight, *Fusarium* wilt and stunt virus (pea leaf roll) are the major diseases. Application of insecticides and fungicides is effective but often uneconomic. For harvest, entire plants are pulled up and threshed, tall erect cultivars may be combine-harvested. Yield averages 600–700 kg/ha, but may be doubled with proper care and conditions. In Burma, yield varies around 630 kg/ha. Storage suffers from bruchid attack; hygiene is required.

Genetic resources and breeding ICRISAT near Hyderabad, India and ICARDA near Aleppo, Syria, and many national institutions conserve large collections of germplasm. The chickpea has 42 wild relatives, of which 8 are annual and range from the Mediterranean Area to Central Asia. Only *Cicer reticulatum* Ladiz. (= *C. arietinum* var. *reticulatum* (Ladiz.) Cubero & Moreno) produces viable hybrids with chickpea. Breeding is directed towards high-yielding disease-resistant cultivars. Yield potential is high, 2000–4000 kg/ha, but is not achieved under poor conditions.

Prospects In general, chickpeas are pushed to marginal conditions. Despite its adaptability, average yields have not increased. In Indonesia and the Philippines, the crop is under test. Chickpea is suitable for cool post-monsoon seasons. Crops in the rainy season produce vegetative matter suitable for fodder. To produce seed in South-East Asia, the cultivars from Sudan and Egypt may prove valuable.

Literature |1| Saxena, M.C. & Singh, K.B. (Editors), 1987. The chickpea. CAB International, Wallingford, Oxon, UK. and ICARDA, Aleppo, Syria. 409 pp. |2| Singh, K.B. & van der Maesen, L.J.G., 1977. Chickpea bibliography 1930 to 1974. 3146 refs.

L.J.G. van der Maesen

Glycine max (L.) Merr.

An interpretation of Rumphius's Herbarium amboinense 274 (1917).

LEGUMINOSAE

$2n = 40$

Synonyms *Phaseolus max* L. (1753), *Glycine hispida* (Moench) Maxim. (1873), *Soja max* (L.) Piper (1914).

Vernacular names Soya bean (En). Soybean (Am). Soia, soja, pois oléagineux de Chine (Fr). Indonesia: kedelai, kacang Jepun, kacang bulu. Malaysia: kacang soya, kacang bulu rimau, kacang Jepun. The Philippines: utau, soybean, balatong. Burma: lasi, pengapi, peryatpym. Cambodia: sândaèk sieng, sândaèk an gen sar. Laos: thwàx khôn, thwàx tê. Thailand: thua lueang, thua phra lueang, thua rae. Vietnam: dâu tuong, dâu nành, quantan.

Origin and geographic distribution Soya bean originated as a domesticate in the eastern half of northern China around the 11th Century BC. From there, it spread to Manchuria, Korea, Japan and the Soviet Union where the centuries-long process of domestication took place. Soya beans were mentioned in Japanese literature around 712 AD. Soya bean was introduced to Korea between 30 BC and 70 AD. In 1765, Samuel Bowen introduced soya bean to the United States from China. Soya beans were introduced from China, Japan and Korea to most of the South and South-East Asian countries through the Silk Route.

Uses Soya beans are used in the preparation of a variety of fresh, fermented and dried food products like milk, tofu, tempeh, miso, yuba, soya sauce and bean sprouts. Soya beans are used not only for food but they serve also as a cure for various diseases and body ailments. Soya beans (preferably black ones) are included in medicines to improve the action of the heart, liver, kidneys, stomach and bowels.

Soya beans are processed to extract oil for food and

for numerous industrial purposes. As an edible oil, it enters the market as salad oil, cooking oil, margarine and shortening. The cake remaining after oil extraction is rich in protein and is predominantly used for feed. Modern uses of soya bean proteins in food include defatted flours and grits, concentrates, isolates, textured flours and textured concentrates.

Production and international trade Total world area of soya bean is 52.6 million ha and production of soya beans is 96 million t. The United States has 48 % of the total area with 56 % of total world production. Brazil ranks second with 10 million ha and a production of 16 million t. China is the leading producer in Asia with 10 million t from 7.5 million ha. Among the South and South-East Asian countries, India and Indonesia were the major producers in 1986 with 1.0 and 1.2 million t from 1.35 and 1.2 million ha, respectively. Thailand produced 334 000 t from 244 000 ha in the same year. The Philippines produced 8000 t from 8000 ha, Vietnam 120 000 t from 14 900 ha. The Asian countries as a whole import 8.0 million t of beans, 1.5 million t of oil and 1.8 million t of soya bean meal annually. Of the beans, Indonesia and Malaysia import 3000 and 2000 t/year, respectively. In the region, only Thailand exports 1000 t of soya beans annually. Asia produces half of its domestic demand; the rest is imported. Even though the area is small and production is lower in Asia than for the leading producers like the United States and Brazil, almost all Asian countries produce and consume soya beans. In 1986, the United States exported 21.6 million t of soya beans. Brazil and Argentina are the next largest exporters of soya beans. Among the importers of soya bean in Asia, the scene is dominated by China, Japan, Taiwan and Korea. Indonesia, Malaysia and the Philippines import it too. The world price of soya bean is around US$ 220 per tonne.
Soya bean is grown in India, Indonesia, Thailand and other Asian countries mainly by smallholders. Estates are rather rare.

Properties Per 100 g edible portion dry soya beans contain: water 10 g, protein 35 g, fat 18 g, carbohydrates 32 g, fibre 4 g, ash 5 g. The energy content averages 1680 kJ/100 g. The range of content of protein and of oil is 35–54 % and 13.3–26.7 %, respectively. Yield of meal from soya beans is 80 % and of oil 18 %, the mass ratio of meal to oil being 4.4 : 1. Soya oil is rich in polyunsaturated fatty acids, especially linoleic acid. It contains no cholesterol. It is rich in vitamin E and a useful by-product of the oil is lecithin. The diet does not need to be supplemented with methionine if rate of intake of soya protein is 6 g/day or more. Heat-labile antinutritional factors of soya bean are trypsin inhibitors, haemagglutinins, a goitrogen, antivitamins and phytates, and heat-stable ones are saponins, oestrogens, flatulence factors and lysinoalanine. The weight of soya beans is 5–70 g/100 seeds. Small beans (less than 10 g/100 seeds) are used for culturing bean sprouts and larger beans (> 30 g/100 seeds) are used directly as vegetable. Soya beans serve as a major source of Ca and vitamins B-1 and C in the Oriental diet. Protein concentrates and isolates of soya bean have a protein content of 67 % and 93 %, respectively.

Description Usually an erect bushy annual herb, 0.2–1.5 m tall, sometimes viny, brownish or greyish pubescent. Tap-root branched, up to 2 m long, side-roots spreading horizontally to a distance of up to 2.5 m in the upper 10–15 cm; in the presence of *Rhizobium japonicum* bacteria, root nodules are formed. Stem branched or unbranched, becoming woody. Leaves alternate, trifoliolate, glabrous to pubescent; petiole long, especially in lower leaves; leaflets ovate to lanceolate, 3–10 cm × 2–6 cm, entire, base rounded, apex

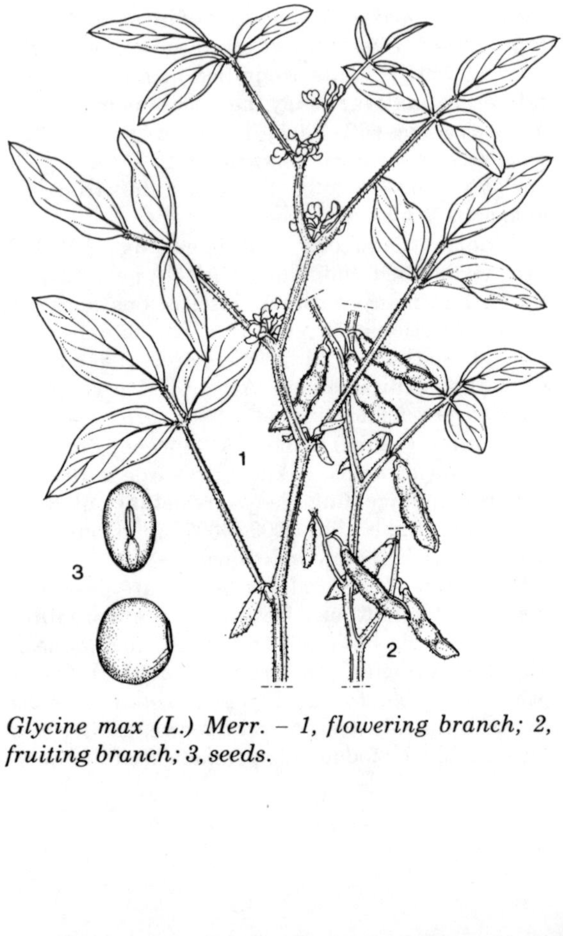

Glycine max (L.) Merr. – 1, flowering branch; 2, fruiting branch; 3, seeds.

acute to obtuse. Inflorescences axillary or terminal racemes with 3–30 flowers; flowers small, papilionaceous, purple or white; calyx tubular, with 2 upper and 3 lower unequal lobes, persistent; keel shorter than the wings, not fused along the sutures; stamens 10, diadelphous; style curved with capitate stigma. Pods slightly curved and usually compressed, 3–15 cm × 1 cm, dehiscent, containing (1–)2–3(–5) seeds; seeds usually globose, yellow, green, brown or black, or blotched and mottled in combinations of those colours; hilum small. Seedlings with epigeal germination; primary leaves simple and opposite.

Growth and development Within 5–15 days of sowing, the seedlings emerge. After 3–10 days, the cotyledons unfold and in the next 3–10 days the first trifoliolate leaf unfolds. Above the primary leaf, 8–24 nodes may develop. Flowering starts from 25 to more than 150 days after sowing, depending on daylength, temperature and cultivar. Flowering can take 1–15 days. Pod formation takes 7–15 days; seed filling 11–20 days; ripening to harvest 7–15 days.

The cycle from sowing to maturity varies from 65 to more than 150 days. Soya bean is a quantitative short-day plant and so development to maturity is usually shorter with short days than long days. At higher altitudes with lower temperatures, flowers are usually cleistogamous. Soya bean is normally self-pollinated and completely self-fertile with less than 1 % cross-pollination. The number of pods per plant varies from a few to more than 1000.

Other botanical information Numerous cultivars are recognized in the Far East that vary in time to maturity, height, plant type, size, colour, content of oil and protein in the seed, and uses to which they are put. For oil production, yellow seeds are preferred. For immature seeds to be used as a vegetable, forms with large yellow or green seeds are preferred. Hay and fodder cultivars usually have brown or black seeds and the plants often twine. Early-maturing Indonesian cultivars include 'Lokon', 'Guntur' and 'Tidar'; other widely used Indonesian cultivars are 'Orba', 'Wilis', 'Kerinci', 'Dempo' and 'Galunggung'. 'SJ4' and 'SJ5' are popular cultivars in Thailand. A narrow-leaflet cultivar, 'Sukothai 1', is high-yielding, resistant to bacterial pustule and widely grown during the rainy season in Thailand. Another early-maturing high-yielding cultivar, 'Nakhonsawan 1', is also popular in Thailand. The following cultivars are widely grown in India: 'Ankur', 'Alankar', 'Shilajeet', 'Gaurav', 'Durga', 'KHSb-2' and 'PB-1'. In Taiwan, 'KS No 10' is a cultivar that has narrow leaflets and that is resistant to bacterial pustule. 'KS No 1' is a newly released vegetable cultivar.

Thirteen maturity groups (MG) are recognized in the United States and Canada. The groups 000, 00 and 0 are early and adapted to the extreme northernmost latitudes; MG IX and X are adapted to equatorial latitudes. Although there are wet-season and dry-season soya beans, no definite classification has been based on wet and dry seasons. There is considerable genetic diversity in the time needed to reach maturity under short-day conditions in the world collection of soya bean germplasm.

Ecology Soya bean is grown from the Equator to latitude 55 °N or 55 °S, and from below sea level to altitudes close to 2000 m. Soya bean is a short-day plant. Some cultivars respond as quantitative short-day plants and some are almost completely insensitive to photoperiod. The response to photoperiod is readily modified by temperature. Differences in photoperiod, temperature and the relative sensitivity of genotypes to those factors dictate the rates and durations of phenological development of soya bean, whether they are grown in temperate areas or in the tropics. Temperatures below 21 °C and above 32 °C can reduce floral initiation and pod set. Extreme temperatures above 40 °C are detrimental for seed production. If water is available, soya beans can be grown throughout the year in the tropics and subtropics.

Uptake of water by soya bean is as much as 7.6 mm/day, requiring 500 mm in a season for a good crop. Drought stress during flowering reduces pod set but it reduces yield more at pod-filling stage than at flowering stage. Soya beans can tolerate brief waterlogging but weathering of seed is a serious problem in the rainy season.

Soya beans are sensitive to low pH. In acid soils, liming is essential to raise the pH to 6.0 or 6.5 and to obtain optimum yield. Toxicosis with Mn, Fe and Al is common with low pH, and deficiency in Mn and Fe with high pH. Cultivars tolerant to Fe deficiency are available.

Propagation Soya beans are propagated by seed. They are cultivated both as a sole crop and in various intercropping systems with maize, cassava, sorghum, banana, sugar-cane, rubber, oil palm, coconut and fruit-trees. In maize and sorghum, soya beans can be intercropped with two rows. Soya beans are grown on paddy-rice bunds too. Soya beans are sown without tillage in rice stubble after each harvest in rows with a spacing of 25 cm × 25 cm or 20 cm × 20 cm. In tilled fields, soya beans are sown in rows 40–50 cm apart and within rows the seeds are either drilled or planted 10 cm apart. Within each hill, two to three seeds are planted. Broadcasting of seed after the rice harvest is also practised. In Indonesia, some farmers soak the seed overnight and then sow by broadcasting. Most of the soya bean farmers in South and South-East Asia are smallholders.

Husbandry Weed control is essential and the timing and frequency varies with season, cultivar and locality. Irrigations at flowering and during seed filling are essential to gain optimum yield. More frequent irrigation is needed in sandy well drained soils than in heavy clay soils.

To obtain seeds at 1 t/ha, soya bean plants must absorb nitrogen at 80 kg/ha. About half of that N is provided by symbiotic fixation by *Rhizobium*. The rest should come from the soil or from N fertilizer. The uptake of P and K by soya bean is around 90 and 85 kg/ha, respectively.

Diseases and pests Among the diseases, soya bean rust caused by *Phakopsora pachyrhizi* can reduce yields by as much as 90 %. Other serious diseases in the region are: bacterial pustule (*Xanthomonas campestris* pv. *phaseoli*), downy mildew (*Peronospora manshurica*), anthracnose (*Colletotrichum truncatum* and *Glomerella glycines*), purple seed stain (*Cercospora kikuchii*), pod and stem blight (*Phomopsis sojae* (teliomorph) *Diaporthe phaseolorum* var. *sojae*), soya bean mosaic virus, yellow mosaic virus and various seedling diseases. Among the insect pests, beanflies (also called stem miners) *Melanagromyza sojae*, *M. dolichostigma*, *Ophiomyia centrosematis* and *O. phaseoli* can cause 100 % yield loss. Pod feeders such as stink bugs (e.g. *Nezara viridula*, *Riptortus clavatus*, *R. linearis* and *Piezodorus hybneri*), pod borers (e.g. *Leguminivora glycinivorella*, *Etiella zinckenella*, *E. hobsoni*, *Heliothis armigera* and *Maruca testulalis*), and some defoliators (e.g. *Heliothis armigera*, *Spodoptera litura*, *S. exigua*, *Plusia chalcites*, *Heydelepta indicata*, *Lamprosema indicata*, *Phaedonia inclusa* and *Diacrisia obliqua*), are serious risks in soya bean production. Nematodes can reduce yields too.

However chemical control is the only means of controlling the pests. Socio-economics and availability of resources are major considerations in the application of chemical control methods.

Harvesting Early-maturing cultivars can be harvested for grain 70 days after planting and late maturing cultivars need up to 160 days. The plants are cut near the ground or pulled with their roots when most leaves have aged and turned yellow, and when the pods have turned brown or black. Combine-harvesting is common in developed countries but is not widely adopted in developing countries. Vegetable soya beans are harvested when the pods are still green but the seeds fill the pod.

Yield The average yield of grain soya bean in South-East Asia in 1986 was 1073 kg/ha against 1841 kg/ha for the world. The average yield in Indonesia was 999 kg/ha, whereas in India it was only 741 kg/ha. Thailand had an average yield of 1369 kg/ha and the United States produced 2271 kg/ha. The maximum yield obtained in South-East Asia was more than 5000 kg/ha. The yield of intercropped soya bean was 0.6–2.0 t/ha in trials. The marketable pod yield of vegetable soya bean in Taiwan was 6–8 t/ha.

Handling after harvest The plants with the pods are dried in the sun. They are then trampled by cattle or tractor, to facilitate threshing, which may also be done by beating with sticks. The seeds are winnowed, cleaned and prepared either for store or market. Deterioration of seed in storage is a problem in the tropics and is attributable to inappropriate storage conditions. As a vegetable, whole pods are marketed either fresh or frozen. Sometimes the seeds are shelled from the pods like green peas and marketed. Export of frozen vegetable soya bean is an industry worth many millions of dollars in Taiwan.

Genetic resources Five provinces in China maintain collections with a total of 12 390 accessions. At AVRDC, the collection of germplasm totalled 12 505 accessions, including several duplicates, by the end of 1987. The AVRDC collection includes most of the United States Department of Agriculture collection and germplasm from various Asian countries. The United States Department of Agriculture holds more than 10 000 accessions in their collections in Illinois and Mississippi. India has 5800 accessions. The indigenous soya beans in Indonesia have been collected and maintained at the Malang Research Institute for Food Crops. Japan has an excellent facility storing 3541 accessions of germplasm. South Korea has collected and evaluated 3130 locally cultivated and wild soya beans. Thailand has a duplicate collection from Japan and AVRDC.

Breeding The major breeding objectives are improvement in yield potential, improved disease and pest tolerance, early maturity, good seed quality, improved adaptation to environments, weathering resistance, sometimes promiscuous nodulation, development of tropically adapted cultivars and developing cultivars for purposes such as vegetable soya beans and bean sprouts. Improvement

of soya oil quality and elimination of beany flavour are further objectives. Selection for less sensitivity to photoperiod and temperature and tolerance to major diseases such as soya bean rust, bacterial pustule, downy mildew and virus diseases would enhance yield stability across environments and localities.
Glycine soja (L.) Sieb. & Zucc. is the wild annual relative of *G. max*. It is sometimes considered as a subspecies of soya bean, *G. max* ssp. *soja* (Sieb. & Zucc.) Ohashi. *G. tomentella* Hayata has aneuploids and polyploids and is found in Australia, southern China, Taiwan, the Philippines and Papua New Guinea. *G. tabacina* (Labill.) Benth. with diploid and polyploid types is found in the same countries, and on the Ryukyu Islands, Mariana and other islands in the Pacific Ocean. All the wild relatives have slender viny stems and small dark seeds with a hard seed-coat that must be scarified to germinate.
The following perennial *Glycine* species are diploid and are found in Australia: *G. clandestina* Wendl, *G. clandestina* var. *sericea* Benth., *G. falcata* Benth., *G. latifolia* (Benth.) Newell & Hymowitz, *G. latrobeana* (Meissn.) Benth. and *G. canescens* F.J. Herm. The newly described *G. argyrea* Tindale and *G. cyrtoloba* Tindale might not be the last relatives to be detected in the Australian region. *G. javanica*, once considered as a relative of *G. max*, has been removed from the genus and placed in the genus *Neonotonia*.
Although *G. soja*, *G. tabacina* and *G. tomentella* are readily available, only *G. soja* is readily crossable. *G. canescens* seems to have single-gene resistance to several races of *Phakopsora pachyrhizi* in Australia. Some of the annual wild species like *G. tomentosa* Benth. and some perennial species seem to be adapted to drought, high pH of soil and disease stress.

Prospects World production of soya bean is expected to double in the next two decades. Many farmers in developed and developing countries are harvesting considerably more than the average yields for their countries. Better adapted and improved cultivars are expected to increase and stabilize yield. Increase in the number of scientists and emphasis in national policy on increasing production will lead to higher yields. Collection and exploitation of indigenous germplasm, enhanced resistance to diseases and pests, improvement in nutritional quality and identification of appropriate management practices to exploit the maximum economic yield potential of cultivars should be given priority.

Literature |**1**| Anonymous, 1987. Interim report on the ATA-272 project 1981–1987. Strengthening the Malang Research Institute for Food Crops, Malang, Indonesia. Annex 6: 1–16. |**2**| Dashiell, K.E., Bello, L.L. & Root, W.R., 1987. Breeding soybeans for the tropics. In: Singh, S.R., Rachie, K.O. & Dashiell, K.E. (Editors): Soybeans for the tropics. John Wiley & Sons Ltd, Singapore. pp. 3–16. |**3**| Hume, D.J., Shanmugasundaram, S. & Beversdorf, W.D., 1985. Soybean (Glycine max (L.) Merrill). In: Summerfield, R.J. & Roberts, E.H. (Editors): Grain legume crops. Collins, London. pp. 391–432. |**4**| Hymowitz, T. & Singh, R.J., 1987. Taxonomy and speciation. In: Wilcox, J.R. (Editor): Soybeans: improvement, production and uses. 2nd edition. American Society of Agronomy, Madison, United States. Agronomy Series 16, pp. 23–48. |**5**| Liu, C.P. & Shanmugasundaram, S., 1984. Frozen vegetable soybean industry in Taiwan. In: Ali, Mohammad Md & Siong, L.E. (Editors): Vegetables and ornamentals in the tropics. University of Pertanian, Serdang, Malaysia. pp. 199–212. |**6**| Pendleton, J.W. & Hartwig, E.E., 1973. Management. In: Caldwell, B.E. (Editor): Soya beans: improvement, production and uses. American Society of Agronomy, Madison, United States. Agronomy Series 16, pp. 211–237. |**7**| Shanmugasundaram, S., 1979. Varietal development and germplasm utilization in soybeans. Asian Vegetable Research and Development Center (AVRDC), Shanhua, Taiwan. 36 pp. Technical Bulletin No 13 (78-102). |**8**| Sumarno, 1987. Soya bean breeding for Indonesian cropping system. In: Shanmugasundaram, S., Lastimosa, P. & Llemit, N. (Editors): Soybean varietal improvement. AVRDC. Shanhua, Tainan. pp. 9–13. |**9**| Summerfield, R.J., Shanmugasundaram, S., Roberts, E.H. & Hadley, P., 1986. Soybean adaptation to photo-thermal environments and implications for screening germplasm. In: Shanmugasundaram, S., Sulzberger, E.W. & McLean, B.T. (Editors): Soybean in tropical and subtropical cropping systems. AVRDC. Shanhua, Tainan. pp. 333–352. |**10**| Tschanz, A.T. & Shanmugasundaram, S., 1985. Soybean rust. In: Shibles, R. (Editor): World Soybean Research Conference 3. Westview Press, Boulder, United States. pp. 562–567.

S. Shanmugasundaram & Sumarno

Lablab purpureus (L.) Sweet

Hortus britannicus 1st ed.: 481 (1827).
LEGUMINOSAE
$2n = 22, 24$

Synonyms *Dolichos lablab* L. (1753), *Dolichos bengalensis* Jacq. (1772), *Lablab niger* Medikus (1787).

Vernacular names Lablab, hyacinth bean (En). Dolique lab-lab (Fr). Indonesia: komak. Malaysia: kara-kara, kekara. The Philippines: bátau, itab, pardá. Burma: pe-gyi. Thailand: thua paep. Vietnam: dâu van.

Origin and geographic distribution Lablab is believed to be native to India, South-East Asia or Africa. It is naturalized and cultivated in the tropics and subtropics, particularly in India, South-East Asia, Egypt and the Sudan.

Uses In South-East Asia, lablab is popular as a vegetable, the young fruits eaten boiled like common beans or used in curries, immature green seeds are eaten boiled or roasted, leaves, young shoots and inflorescences are eaten boiled. In other parts of Asia, lablab is predominantly used as a pulse, often as dhal. Sometimes sprouted seeds are sun-dried and stored to use as vegetable. Lablab is used also as fodder, hay, silage, green manure and cover crop.

Properties Per 100 g edible portion immature pods contain: water 82.4 g, protein 4.5 g, fat 0.1 g, carbohydrates 10.0 g, fibre 2.0 g, ash 1.0 g. The energy content averages 1260 kJ/100 g. The content of protein in mature seeds is normally 21–29 g/100 g. Fully ripe Indian seeds contain per 100 g edible portion: water 9.6 g, protein 24.9 g, fat 0.8 g, carbohydrates 60.1 g, fibre 1.4 g, ash 3.2 g. The energy content averages 1403 kJ/100 g. The presence of a cyanogenic glycoside has been reported in certain cultivars. Seed weight varies between 25 and 50 g/100 seeds.

Description A bushy or a climbing and branching, pubescent herbaceous perennial, often grown as an annual, up to 6 m tall, with a well developed tap-root with many laterals and well developed adventitious roots. Leaves alternate, trifoliolate; leaflets broadly ovate, 5–15 cm × 4–15 cm, entire, subglabrous or soft hairy. Inflorescences stiff axillary racemes with many flowers; peduncle 4–23 cm long, often compressed, glabrescent; rachis 2–24 cm long; flowers arising 1–5 together from tubercles on rachis; pedicels short, square, sparsely pubescent; flowers white, pink, red or purple; stamens diadelphous (9 + 1); ovary sessile, 10 mm long, finely pubescent; style abruptly upturned, 8 mm long; stigma capitate, glandular. Pods variable in shape and colour, flat or inflated, 5–20 cm × 1–5 cm, straight or curved, usually with 3–6 ovoid seeds of varying colour and size.

Lablab purpureus (L.) Sweet – 1, flowering branch with young fruits; 2, fruits; 3, seed.

Growth and development Germination is epigeal and normally takes 5 days. Seed remains viable for 2–3 years and on average 85–95 % germinate. Growth period varies from 75 to 300 days. Improved cultivars start fruiting 60–65 days after sowing and continue for 90–100 days. Early-maturing cultivars that can be grown all year round produce pods 60 days after sowing and continue up to 120 days. Mature seeds are harvested 150–210 days after sowing, depending upon cultivar and time of sowing. In India, short-day cultivars start flowering 42–330 days after sowing, depending on the sowing date. The flowers are mainly cross-pollinated.

Other botanical information The variability of lablab is great; many cultivars exist. Many subclassifications of the species can be found in the literature. Some distinguish subspecies, others varieties. For cultivated plants, the distinction of cultivar (cv.) groups seems most appropriate.

- cv. group Lablab (widely distributed): mature seeds with the long axis at right angles to the suture; pods dehiscent or indehiscent; seeds not longer than $\frac{1}{3}$–$\frac{3}{4}$ of the width of the mature pod;
- cv. group Ensiformis (South-East Asia, East Africa): mature seeds with long axis more or less oblique to the suture, nearly filling the mature pod; pods indehiscent; when young, difficult to distinguish from cv. group Lablab;
- cv. group Bengalensis (South Asia, East Africa): mature seeds with long axis parallel to the suture, more or less filling the mature pod, gibbous dorsally and at base; pods indehiscent.

Ecology Lablab is a short-day plant. It requires high temperatures to grow well (18–30 °C). Minimum temperature for growth is 3 °C. Its frost tolerance is low; light frosts damage the leaves but do not kill the plants. It prefers rainfall at 750–2500 mm/year. Once established (2–3 months after sowing), lablab is drought-tolerant. It has a deep root system which can make use of residual soil moisture. It is reported to grow in areas with rainfall at 200–2500 mm/year. Plants do not tolerate standing brackish water or waterlogging. In India and Burma, the plants are often grown on exposed sandy river banks. Provided drainage is good, the plant is extremely tolerant of soil texture, growing in deep sands to heavy clays, pH ranging from 5–7.8. Lablab prefers the lower altitudes but is grown as a dry-land crop up to 2000 m in the tropics.

Propagation Propagation is by seed at 7–10 kg/ha, up to 5 cm deep in a preferably well prepared seed-bed. Lablab can establish itself after being broadcast into roughly ploughed land, if the seed is covered to some extent. As a field crop, lablab is usually sown in rows, either as a sole crop (rows 1 m apart) or intercropped with a cereal, e.g. with maize, rows 80 cm apart. In India, it is invariably grown with *Eleusine coracana*. Inoculation with cowpea-type *Rhizobium* strains is advised, if lablab has not been grown recently. Lablab is only grown by smallholders.

Husbandry Weed control during the early stages of growth may be necessary. Recently the use of pre-emergence herbicides such as chloramben, chlorthal dephenamid, trifluralin and dinoseb has been suggested. Generally no fertilizer is applied on fertile soils. In poor sandy soils, molybdenized superphosphate at 250–500 kg/ha and some potash is advised. In India, garden cultivars are heavily manured, frequently irrigated and supported for climbing. They are sown in pits and thinned to 4 vines after 1 month. Lablab is sometimes grown as a cover crop in rotation with sorghum and cotton. In orchards, lablab forms a good organic mulch when cut often. It can also produce a good yield of hay, which is easily cured through its low moisture content. It also produces good silage, especially when mixed with sorghum.

Diseases and pests Pod-boring larvae are the most serious pests of the lablab bean. *Adisura atkinsoni* is particularly troublesome. That pest has been controlled experimentally by strain HB-III of *Bacterium cereus* var. *thuringensis*. In addition, the gram caterpillar, *Heliothis armigera*, the plume moth, *Exelastis atomosa*, and the spotted podborer, *Maruca testulalis*, are of considerable economic significance. Insect infestation during storage is particularly caused by bruchid beetles, *Callosobruchus* spp., which also attack the crop in the field. Anthracnose, caused by *Colletotrichum lindemuthianum*, can cause serious crop losses; spraying with zineb or captan is reported to give reasonable control. Leaf-spot, caused by *Cercospora dolichi*, and powdery mildew, *Leveillula taurica* var. *macrospora*, may also be troublesome and are controlled by spraying with Bordeaux mixture.

Harvesting The green pods are picked by hand when they have reached a reasonable size, usually when the seeds are three-quarters ripe. They are generally picked from the plants at intervals of 3–4 days, cleaned and graded for size, before being packed in baskets for the market. In many cultivars, the pods mature in succession on the stem and shatter once they are ripe. For seed production, the pods are frequently picked by hand as soon as they ripe, until the plants reach full maturity and the major proportion of the remaining pods has ripened. At that stage, the entire plant is cut close to the ground with a sickle and the vines left to dry for a few days before threshing.

Yield The average yield of green pods is 2600–4500 kg/ha, and of seed is 450 kg/ha if grown as intercrop and up to 1460 kg/ha in sole cropping. Fodder yields are 25–40 t/ha.

Handling after harvest Storage at 0–2 °C and a relative humidity of 85–90 % is reported to extend the shelf-life of the green pods to a maximum of 21 days and of the shelled fresh beans up to 7 days. After drying and cleaning, mature seeds are stored; usually earthenware or metallic containers are used in South-East Asia, and the seeds may be covered with a 5-cm protective layer of sand. Harvesting pods as soon as the seed is ripe reduces bruchid infestation; reduction of the moisture content to below 10 % is more effective.

Genetic resources There is wide genetic variation in the tropics and subtropics. More than 250 lines, both indigenous and exotic, are maintained and catalogued at the University of Agricultural Sciences of Bangalore, India. Small collections are available at TNAU (Tamil Nadu Agricultural University), Coimbatore and IIHR (Indian Institute of Horticultural Research), Bangalore. No attempts have been made so far to collect and catalogue the germplasm elsewhere in the world. In Australia and New Zealand, only fodder types are maintained.

Breeding Most of the improvement work is concentrated in India. The local landraces are of long duration, photosensitive and low-yielding (500–600 kg/ha). The main breeding objective is higher yield. With germplasm collections, India is trying to obtain bushy short-duration day-neutral disease-resistant cultivars.

Prospects As well as being a pulse, lablab is becoming more relevant for green manure, as a cover crop, for pasture and for fodder. Its drought resistance and its suitability for all types of soils gives it a wide range of options. Research priorities are: collection of all cultivars (germplasm); identification of suitable *Rhizobium* strains; investigation of its value as an intercrop with food and forage crops; further evaluation of the nutritional value of the seeds for man and animal.

Literature |1| Duke, J.A., 1981. Handbook of legumes of world economic importance. Plenum Press, New York and London. pp. 102–106. |2| Shivashankar, G., Vishwanatha, S.R., Manjunath, A. & Chandrappa, H.M., 1971. Inheritance studies and breeding in Dolichos. Proceedings of the International Symposium on Subtropical and Tropical Horticulture, February 1972. Bangalore, India. |3| Skerman, P.J., 1977. Tropical forage legumes. FAO, Rome. Plant Production and Protection Series No 2. pp. 314–322. |4| von Schaaffhausen, R., 1963. Dolichos lablab or hyacinth bean: its uses for feed, food and soil improvement. Economic Botany 17:146–153.

G. Shivashankar & R.S. Kulkarni

Lathyrus sativus L.

Species Plantarum 2: 730 (1753).
LEGUMINOSAE
$2n = 14$

Vernacular names Grass pea (En). Gesse blanche (Fr). Burma: pèsali.

Origin and geographic distribution Native to southern Europe and western Asia. Widely grown in Iran, Middle East, parts of Africa and South America, and especially in India.

Uses In India, seeds are eaten in times of food shortages (white seeds preferred). The seeds may be boiled and eaten as a pulse, split and used as dhal, or ground into a flour and made into unleavened bread, paste balls or curries. Young leaves and fruits are eaten as a vegetable too.

It is cultivated as a forage crop and for green manure too. As fodder, the plants can be eaten green or as hay. They are not suited for silage.

Oil from the seeds is medicinally used as a powerful cathartic.

Production and international trade In India, 1.6 million ha is under grass pea, with seed production 0.5 million t/year. The crop is consumed locally and does not enter international trade.

Properties Per 100 g edible portion seeds contain: water 10 g, protein 25 g, fat 1 g, carbohydrates 61 g, fibre 1.5 g, ash 3 g. The energy content averages 1231 kJ/100 g. Contents in hay are: water 14.6 %, protein 9.9 %, fat 1.9 %, fibre 36.5 %, nitrogen-free extract 31.0 %, ash 6.1 %. Seed weight varies between 5–7 g/100 seeds.

People and livestock consuming grass pea as the principle article of diet for months long develop a paralytic disease called lathyrism. Seeds soaked in water for 24 hours before cooking are not toxic. The poison works in various parts of the central nervous system, with great impairment of muscular activity, especially in the lower limbs.

Botany A much-branched, suberect, straggling or climbing annual, with winged quadrangular stems up to 90 cm long. Leaves 2- or 4-foliate; leaflets narrowly elliptic-oblong, 3–6.5 mm × 3–11 mm. Flowers reddish-purple, pink, blue or white. Pods oblong, up to 4 cm long, upper margin 2-winged. Seeds wedge-shaped, diameter 4–7 mm, grey to brown, marbled or light green. Germination hypogeal. Plants are self-fertile. Classification is based on colour of flowers, markings on pods and size and colour of seeds.

Ecology Grass pea is grown as a cold season crop in India. It thrives in areas with average temperature 10–25 °C. It is drought-resistant and is grown successfully with average rainfall of 380–650 mm/year. It tolerates waterlogging and poor soils. In India, grown up to altitude 1200 m.

Agronomy Grass pea is propagated by seed. Seeds are broadcast or sown in furrows 3 cm apart. In a well prepared field, the crop comes up as a thick clove mass over the entire surface and

Lathyrus sativus L. – *1, flowering and fruiting branch; 2, seeds.*

smothers out weeds. Frequently the crop receives virtually no attention after sowing.
Plants are sole-cropped or intercropped, for instance with rice in India. Seed rate is 45–60 kg/ha for sole crop, 34 kg/ha when intercropped. It has no serious diseases nor pests. Ripe seeds can be harvested 4–6 months after sowing. Average yield of seed is 1000–1500 kg/ha.

Genetic resources and breeding Primitive forms of *L. sativus* can be found in and around north-western India. In the Mediterranean, forms with large bright seeds are present. *L. cicera* L. from the Mediterranean and western Asia is a close relative. To solve the problem of lathyrism, low-toxin cultivars are needed.

Prospects Grass pea will remain useful as a crop for extremely dry and poor soils and as a rescue crop when other crops have failed. In South-East Asia, it is possibly interesting for dry areas. More research is needed on effective methods of detoxifying the seeds in the country without reducing their nutritive value.

Literature |1| Duke, J.A., 1981. Handbook of legumes of world economic importance. Plenum Press, New York and London. pp. 107–110. |2| Kay, D.E., 1979. Food legumes. Tropical Products Institute, London. Crop and Product Digest No 3, pp. 115–123.

P.C.M. Jansen

Lens culinaris Medikus

Vorlesungen der churpfälzischen physikalisch-ökonomischen Gesellschaft 2: 361 (1787).

LEGUMINOSAE

$2n = 14$

Synonyms *Lens esculenta* Moench (1794).

Vernacular names Lentil (En). Lentille (Fr). Cambodia: lânti. Thailand: thua daeng.

Origin and geographic distribution Lentil is one of the oldest pulse crops and of ancient cultivation in western Asia, Egypt and southern Europe. It is widely cultivated in temperate and subtropical regions and in the tropics at higher altitudes, especially in the Indian Subcontinent.

Uses In India, split seeds (dhal) are used in soups, young pods are used as a vegetable. Flour from ground seeds, mixed with flour from cereals, is considered a nourishing food. In parts of India, the whole seed is often eaten salted and fried.
The whole plant, green or dry, makes excellent fodder and green manure. Seeds are a source of commercial starch for textile and printing industries. Lentils are supposed to remedy constipation and other intestinal afflictions. In India, lentils are poulticed onto slow-healing sores.
Seed production and international trade World area under cultivation is 1.8 million ha with a seed production of 1.1 million t/year. India is the major producer with 428 000 t/year from 925 000 ha.
In the Near East, production is mainly for local use. Turkey is the largest exporter with a production of 200 000 t/year from 200 000 ha.

Properties Per 100 g edible portion dry seeds contain: water 11 g, protein 24 g, fat 1.8 g, carbohydrates 59 g, fibre 1.8 g and ash 2.2 g. The energy content averages 1450 kJ/100 g. Seed weight varies between 1 and 9 g/100 seeds. Lentils are considered to be the most easily digested of grain legumes.

Botany An annual, much-branched herb, with square stems up to 50(–75) cm tall. Leaves with 3–8 pairs of leaflets, usually ending in a tendril; leaflets narrowly obovate-elliptic, 8–18 mm × 2–5 mm. Flowers small, blue to white, 1–4 together in axillary racemes. Pods rhomboidal, laterally compressed, 6–20 mm × 4–12 mm, with 1–2 seeds. Seeds lens-shaped, diameter 3–9 mm, green,

Lens culinaris Medikus – 1, flowering and fruiting branch; 2, seeds.

brownish green or light red speckled with black. Seedling with hypogeal germination. For germination, a minimum temperature of 15 °C is required; optimum temperature is 18–21 °C.

Flowering starts 6–7 weeks after sowing. Harvest is possible 3–4 months after sowing. Lentils are usually self-fertilized but up to 0.8 % are cross-pollinated. Lentils are nodulated by *Rhizobium leguminosarum*.

L. culinaris is divided into 3 subspecies.

- ssp. *culinaris*, to which most cultivars belong: stipules entire, lanceolate; leaflets 6–16; peduncles aristate, shorter or subequal to the rachis; calyx teeth smaller, subequal or greater than corolla;
- ssp. *nigricans* (M. Bieb.) Thell.: stipules toothed; leaflets up to 10; peduncles aristate, longer or subequal to the rachis; calyx teeth subequal or greater than corolla;
- ssp. *orientalis* (Boiss.) Ponert: stipules entire; leaflets 6–12; peduncles non-aristate, usually awned, longer or subequal to the rachis; calyx teeth smaller or subequal than corolla.

Ecology Optimum mean temperature for yield is 24 °C. Annual mean temperatures of 6–27 °C are tolerated. Average moisture requirement is 750 mm/year but rainfall of 280–2430 mm/year is tolerated. Frost and temperatures above 27 °C affect growth seriously.

Lentils tolerate a pH of 4.5–8.2, optimum pH is 5.5–7.

In the tropics, lentils are cultivated up to altitude 3800 m but the crop is not suited to the hot humid tropics. They are less damaged by drought than by waterlogging. They thrive on a wide range of soils from light loams and alluvial soils to black cotton soils, but best on clay soils.

Lentils are quantitative long-day plants, some cultivars tending to be day-neutral.

Agronomy Lentil is propagated by seed. The field should be ploughed and harrowed to a fine texture. Seeds may be either broadcast or sown in drills at a rate of 25–90 kg/ha, in rows 60–90 cm apart and plants 17–30 cm apart, if sole-cropped.

Lentil is, however, seldom sole-cropped. In India, it is grown intercropped, for instance with rice, barley, castor and mustard, at seed rates of 11–13 kg/ha. In other areas, it is often alternated with wheat or grown as a winter crop.

Weed control is necessary. Two weedings usually suffice for weed control during establishment. Otherwise pre-emergence herbicides are applied. In the Middle and Near East, broomrapes (*Orobanche* spp.) parasitize the crop.

The crop is usually rain-fed. Fertilizer is not usually applied but some fertilizer at planting and at flowering may be beneficial.

Lentils are harvested when the pods are golden-yellow and the lower ones are still firm (to prevent shattering). At harvest, plants are cut to the ground, dried for 7–10 days and then combined or threshed. Yields of seeds are 340–675 kg/ha in dry cultivation, may increase to 1700 kg/ha with irrigation and have reached 3000 kg/ha in trials. Lentils require careful cleaning for seed as it is often badly contaminated with weed seed. Seeds should be dried to a moisture content of 11–14 %. Below 11 %, shells are hard and seeds tend to break.

Serious diseases are the wilt *Fusarium oxysporum* f. sp. *lentis*, the blight *Ascochyta lentis* and the rust *Uromyces fabae*. Serious pests are caused by pea leaf weevils (*Sitonia* spp.) and the gram caterpillar (*Heliothis obsoleta*).

Genetic resources and breeding ICARDA, Aleppo, Syria, the world centre for improvement of lentil, has 6000 accessions from 47 countries.

Breeding objectives are adaptation to mechanized harvest (erect types, resistance to lodging and to fruit dehiscence), reducing hazards from pests, diseases and weeds, increasing yield and stability of yield, and adaptation to short growing season with limited moisture resources.

Wild relatives are urgently needed to improve the tolerance to environmental stresses. Wild relatives are:

- *L. culinaris* ssp. *orientalis* (formerly *Lens orientalis* (Boiss.) Handel- Mazzetti), occurring in the Near East, possibly the progenitor of the cultivated lentil;
- *L. culinaris* ssp. *nigricans* (formerly *Lens nigricans* (M. Bieb.) Godron);
- *L. ervoides* (Brign.) Grande, tropical Africa and Mediterranean Region;
- *L. montbrettii* Fisch. & Mey. (now transferred to the genus *Vicia*, $2n = 12$).

Prospects The prospects are good for this most nutritious pulse. Especially in North and South America, production is increasing because farmers wish to include legumes in rotations with cereals.

Being a crop for marginal areas, there is an urgent need for well adapted short-duration cultivars. In South-East Asia, lentil might be an interesting crop for dry poor high-altitude areas.

Literature |**1**| Muehlbauer, F.J., Cubero, J.I. & Summerfield, R.J.,1985. Lentil. In: Summerfield, R.J. & Roberts, E.H. (Editors): Grain legume crops. Collins, London. p. 266–311. |**2**| Webb, C. & Hawtin, G.C. (Editors), 1981. Lentils. Commonwealth Agricultural Bureau, Farnham Royal, Bucks., GB.

P.C.M. Jansen

Macrotyloma uniflorum (Lam.) Verdc.

Kew Bulletin 24: 322, 401 (1970).

LEGUMINOSAE

$2n = 20, 22, 24$.

Synonyms *Dolichos uniflorus* Lam. (1786), *Dolichos biflorus* auct. mult., non L.

Vernacular names Horse gram (En). Grain de cheval (Fr). Indonesia: kekara. Burma: pè-bi-zât.

Origin and geographic distribution Horse gram is native to the Old World tropics. It is extensively cultivated in the drier areas of Australia, Burma, Sri Lanka, the Himalayas, Africa and America, and especially in India.

Uses As a pulse, the seeds are eaten poached, boiled or fried, whole or ground. In Burma, boiled seeds are pounded with salt and fermented, which results in a product similar to soya sauce.

All parts are used as fodder and as green manure. Its use as fodder is gaining ground in Australia and South-East Asia.

In traditional medicine, the seeds are said to be used as an astringent, diuretic and tonic.

Production and international trade For the period 1968–1972, production in India has been estimated to be 377 000 t/year. Most is produced for domestic consumption. International trade is of no importance.

Properties Per 100 g edible portion seeds contain: water 8–12 g, protein 22–25 g, fat 0.5–5 g, carbohydrates 57–60 g, fibre 5 g, ash 3 g. Energy content averages 1400 kJ/100 g. Seed weight is about 1.5 g/100 seeds.

Botany A low-growing, slender, suberect annual or perennial herb with slightly twining downy stems and branches, 30–60 cm high. Leaves trifoliolate; leaflets broadly ovate, 2.5–5 cm long. Flowers 1–3 together on short axillary peduncles; corolla pale yellow. Pod linear, 3–5 cm long, recurved, beaked, downy, dehiscent, with 5–7 seeds. Seeds small, rhomboid, 3–6 mm long, flattened, light red, brown, black or mottled. Horse gram is self-fertile.

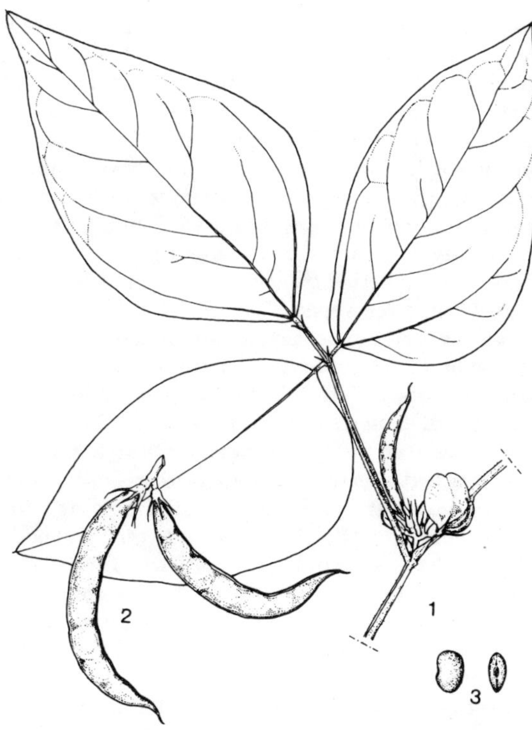

Macrotyloma uniflorum (Lam.) Verdc. – 1, branch with flowers and young fruit; 2, fruits; 3, seeds.

Four varieties are distinguished.
- var. *uniflorum*: pods 6–8 mm wide. Native of India. Widely cultivated in the tropics as cover and forage crop;
- var. *stenocarpum* (Brenan) Verdc.: pods 4–5.5 mm wide; leaflets pubescent; sutures of pods not warted. Wild plant in *Acacia* bushland and thicket in Africa and India;
- var. *verrucosum* Verdc.: pods 4–5.5 mm wide; leaflets pubescent; sutures of pods obscurely to markedly warted. Wild plant in grassland in Kenya, Tanzania and Mozambique;
- var. *benadirianum* (Chiov.) Verdc.: pods 4–5.5 mm wide; leaflets densely velvety; sutures of pods slightly warted. Wild plant in sand dunes of Somalia and Kenya.

Ecology Horse gram is a short-day plant. It requires an average temperature of 20–30 °C and is completely intolerant of frost. It is grown as a dry-land crop in areas with rainfall less than 900 mm/year; in high rainfall areas, it is sown in the dry season. It does not tolerate waterlogging. It does well on light sandy soils, red loams, black cotton loams and stony and gravelly upland soils with pH 5–7.5. In India, it is grown up to altitude 1800 m.

Agronomy Horse gram is propagated by seed. Seeds are broadcast or sown 1–2.5 cm deep in rows 90 cm apart at a rate of 25–45 kg/ha. It does best in a well prepared seed-bed but will establish in soil that has hardly been disturbed.
It is usually grown as a sole crop but intercropping is possible, for instance with *Guizotia*, *Eleusine*, *Lens*, castor beans, groundnuts or maize. It can also easily be established in natural pastures. The crop requires little attention. Its vigorous growth tends to smother weeds.
The most serious diseases in India are a root-rot caused by *Rhizoctonia* sp. and anthracnose (black sunken spots on stems, leaves, seeds and pods) caused by *Glomerella lindemuthianum*. The gram leaf caterpillar *Azazia rubricans* is destructive.
The crop matures in 4–6 months for seed. For forage, harvest can start 6 weeks after sowing. For seed production, the plants are cut when the pods are dry, stacked and dried in the sun for 7 days. Plants are then threshed by flailing with sticks, by stone rollers or in India by treading with oxen. Seed requires thorough winnowing and sieving as weed content is usually high.
For seeds, yield averages 130–900 kg/ha in India, 1120–2240 kg/ha in Australia; for green forage expressed as dry matter, 5–14 t/ha in India, 4.4 t/ha in Australia.

Genetic resources and breeding The cultivated crop is usually a mixture of several landraces with different seed colours and periods of maturity.
Breeding is focused on day-neutral cultivars with high yield potential. Wild relatives are the three varieties mentioned under Botany.

Prospects For dry areas, horse gram is an interesting crop because of its rapid growth, its heavy seeding habit and its drought tolerance. In Australia, its use as fodder is restricted because of its short life and its loss of leaf in autumn and winter.

Literature 1 Duke, J.A., 1981. Handbook of legumes of world economic importance. Plenum Press, New York and London. p. 146–149. |2| Kay, D.E., 1979. Food legumes. Tropical Products Institute, London. Crop and Product Digest No 3, p. 177–183.

P.C.M. Jansen

Phaseolus acutifolius A. Gray

Plantae wrightianae texano-neo-mexicanae 1: 43 (1852).
LEGUMINOSAE
$2n = 22$

Synonyms *Phaseolus tenuifolius* Woot. & Standley (1913).

Vernacular names Tepary bean (En).

Origin and geographic distribution The tepary bean is native to the south-western United States and Mexico. It has been introduced and is cultivated also in Africa, Asia and Australia. Cultivation is most important in Mexico and Arizona (United States).

Uses Tepary beans are used mainly as dry shelled beans, in Mexico and Uganda to prepare a soup. In the United States, the plants have been tried as a hay and cover crop. In areas of poor rainfall, tepary bean could be used as a catch crop where a rapid food supply is needed. On storage, the dry seeds become very hard and take a long time to cook.

Production and international trade Tepary beans are mainly grown in Mexico and Arizona. In tropical Africa, they are grown year-round as a food crop. No production data are available.

Properties Per 100 g edible portion tepary beans contain: water 10 g, protein 22–25 g, fat 1.5 g, carbohydrates 60–65 g, fibre 3–4 g, ash 4 g. The energy content averages 1588 kJ/100 g. Seed weight is 15 g/100 seeds.

Botany A suberect annual herb up to 30 cm high,

bushy on poor land, otherwise recumbent, spreading or twining to 2 m long. Leaves trifoliolate; leaflets ovate, 4–8 cm × 2–5 cm. Inflorescences axillary with 2–5 white or pale lilac flowers. Pods straight or slightly curved, compressed, 5–9 cm × 0.8–1.3 cm, with 2–7 seeds, rimmed on margins, sharp beaked, hairy when young. Seeds roundish to oblong, 8 mm × 6 mm, white, yellow, brown or deep violet, sometimes variously flecked. The seeds absorb water easily. Germination is epigeal. First pair of leaves is simple. Tepary bean is presumably self-fertilized. It gives a crop in as soon as 2 months but some cultivars need up to 4 months.

Two varieties are distinguished:
- var. *acutifolius*, the wild forms; seeds small and greyish; leaflets very variable from ovate to linear; northern Mexico, southern United States;
- var. *latifolius* Freeman, the cultivated bean; seeds larger and variously coloured; preferably grouped as cv. group.

Ecology Tepary beans are short-day or day-neutral plants. They withstand heat and dry atmosphere and are drought-resistant. They are susceptible to waterlogging and frost, and are not suited to the wet tropics.

Temperature may not drop below 8 °C. Annual mean temperature should be 17–26 °C. It can grow in areas with a rainfall of 500–600 mm/year, in areas up to 1700 mm/year. It can be grown on poor shallow soils of pH 5–7. In areas with average rainfall of 1000 mm/year or more, vegetative growth is usually excessive at the expense of seed yield.

Agronomy Tepary bean is propagated by seed. Seeds may be broadcast at rate of 28–33 kg/ha or drilled 2–10 cm deep, in rows of 90 cm × 45 cm, at a rate of 11–17 kg/ha; sometimes they are planted on hills 45 cm apart with 2–4 seeds per hill. When grown for hay, it is usually sown at more than 60 kg/ha.

Rhizobium strains that nodulate Lima beans and *Canavalia* species induce nitrogen fixation in tepary beans too.

The crop may be irrigated, though not usually. It is sensitive to salty water and soils. It is often grown as a catch crop and requires little weeding, if grown as an end-of-season crop. Tepary bean responds favourably to N fertilizer and to K at a rate of K_2O of 75 kg/ha.

Tepary bean is practically disease-free under savanna conditions. In Burma, it was susceptible to attack from a root rot, *Rhizoctonia* sp.

As soon as the first pods start to ripen, whole plants are pulled up, dried and threshed. Yield averages 500–800 kg/ha under dry farming, 900–1700 kg/ha under irrigation. Yields of dry beans up to 5000 kg/ha are reported. Yield of oven-dry hay is 5–10 t/ha.

Prospects Tepary bean is reported to exhibit tolerance to disease, drought, heat, sand, slope and virus. It produces a crop where other beans fail. In South-East Asia, it could be used as a catch crop where a rapid food supply is needed in areas of poor rainfall.

Literature |1| Duke, J.A., 1981. Handbook of legumes of world economic importance. Plenum Press, New York and London. p. 187–189. |2| Kay, D.E., 1979. Food legumes. Tropical Products Institute, London. Crop and Product Digest No 3. p. 371–376.

P.C.M. Jansen

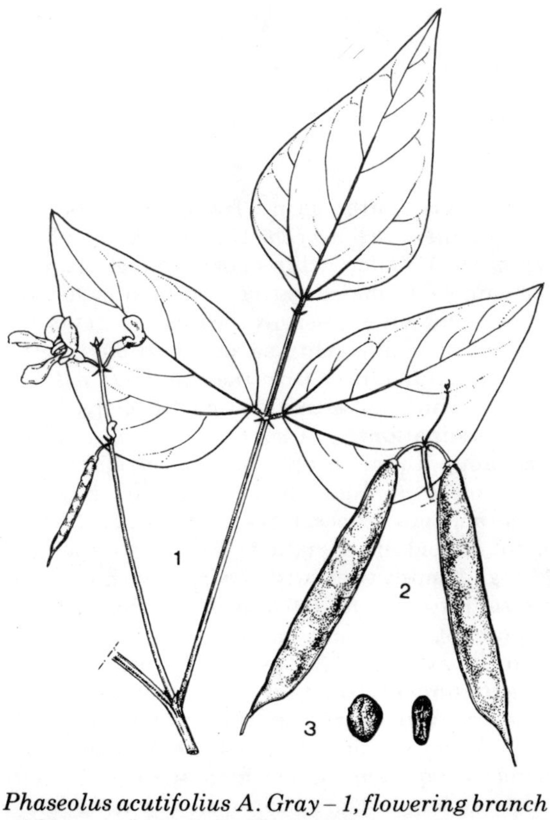

Phaseolus acutifolius A. Gray – 1, flowering branch with young fruit; 2, fruits; 3, seeds.

Phaseolus coccineus L.

Species Plantarum 2: 724 (1753).
LEGUMINOSAE
$2n = 22$

Synonyms *Phaseolus multiflorus* Lam. (1789).

Vernacular names Runner bean (En). Haricot d'Espagne (Fr).

Origin and geographic distribution The runner bean was domesticated in Mexico and perhaps in Guatemala. It is cultivated in highland areas of Africa and South America and in many temperate areas.

Uses Runner beans are used as a green slicing bean and as a pulse. Many recent cultivars have substantial reduction in the fibrous vascular strands of the pod sutures, the stringless runner beans. Preparation is predominantly by boiling. Runner beans are often grown as ornamentals in areas too warm for satisfactory production of pods and seeds.

Production and international trade Production is almost exclusively for local use. Commercial production of pods is done in Great Britain and Argentina and of butter beans (white-seeded cultivars) in South Africa. Accurate global production figures are not available.

Properties The green pods are good sources of vitamins, protein and dietary fibre. Dried seeds contain per 100 g edible portion: water 12.5 g, protein 20.3 g, fat 1.8 g, carbohydrates 62.0 g, fibre 4.8 g, ash 3.4 g. The energy content averages 1420 kJ/ 100 g. Seed weight is up to 100 g/100 seeds.

Botany A climbing (viny) tuberous perennial or a bushy (dwarf), annual herb. Stems up to 4 m length. Leaves trifoliolate, leaflets up to 15 cm long. Inflorescence axillary or terminal, long, producing many, few-flowered nodes. Petals scarlet, pink or white; keel spirally twisted to accommodate the style. Fruits linear, straight or slightly curved with a prominent beak, up to 40 cm long, fleshy, usually green. Seeds ovoid, variable in size, pink to purple dark mottled or sometimes black, white, cream or brown. Germination hypogeal, 10–14 days after sowing. Seedling with first pair of leaves opposite and simple. Flowering 40–60 days after sowing; flowers open at sunrise and fade at sunset. Pollination is by insects. Pod set is favoured by warm moist conditions and is adversely affected by hot dry environments.

Cropping of green pods starts 3 months after sowing and can be easily sustained for 2–3 months. Mature seed can be harvested 4–5 months after sowing. Dry weather at harvest is essential for

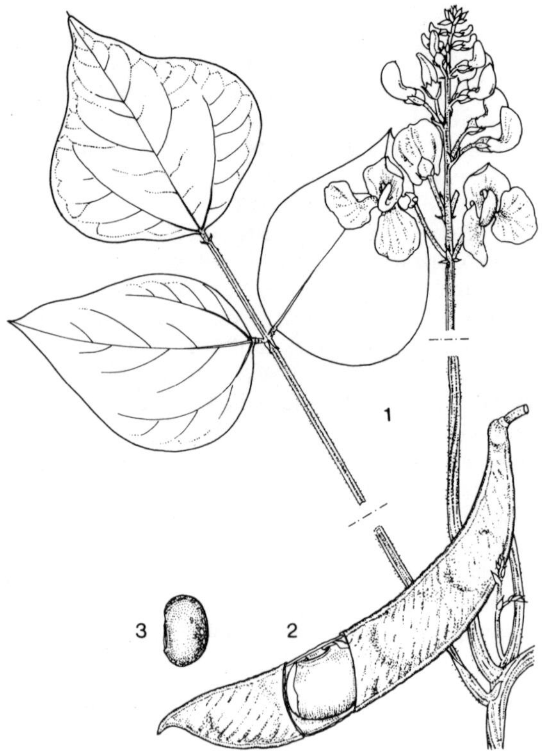

Phaseolus coccineus L. – *1, flowering branch; 2, fruit; 3, seed.*

good-quality butter beans. Dwarf cultivars produce smaller earlier crops than do the climbing types. The productive life-span of runner beans is commonly terminated by frost, which kills the aerial parts but not necessarily the tuberous roots, by excessively cool conditions and short daylength. The main differences from the similar *P. polyanthus* Greenm. are extrorse stigma and hypogeal seed germination.

Ecology Runner bean is most successful in the (sub)tropics at high altitudes of 1500–2000 m, where ambient temperatures are without extremes of hot or cold. Its climatic optimum is temperate. At high temperatures with low relative humidity, pod set is totally inhibited. Waterlogging is detrimental but the crop is more tolerant of cold wet conditions than other *Phaseolus* species.

Agronomy Propagation is by seed. High-quality pods are grown on trellises, poles or other support because they should not touch the soil. Labour and material requirements are high and may inhibit cultivation except on a small scale. Climbing genotypes can yield without support, if leading shoots are pinched out to induce a bushy growth form. Bushy cultivars are generally more satisfactory for seed than for pod production. Planting densities are 50 000–75 000 plants/ha for climbers and double those for bush types, requiring seed to be sown at 75 and 150 kg/ha, respectively. The plant is shallow-rooting and not drought-tolerant. Supplementary irrigation is beneficial. Tillage should be shallow to avoid root damage.

In general, diseases and pests are similar to those in common bean. Bacterial halo blight, *Pseudomonas phaseolicola*, and common blight, *Xanthomonas phaseoli* are tolerated better.

Green pods are ideally harvested when length reaches maximum before the phase of rapid seed development. Crops for dry seed can be pulled or cut when most pods are dry and then allowed to dry out for a few days. Completely dry pods can be threshed. The practice of plucking leaves for a green vegetable is not established but it can be regarded as a potential food resource. Yields of green pods of 10 t/ha and of seeds of 1.5 t/ha are possible.

Genetic resources and breeding Germplasm resources within runner bean are limited. For the improvement of runner bean, the combined genetic resources of *P. vulgaris* and *P. polyanthus* are available. Breeding efforts have been directed to improvement of culinary quality (stringlessness) and pest and disease resistance. Selection to improve cooking quality is promising since seed proteins of runner bean are more polymorphic than of common bean. For dry seed production, improvement of the shoot canopy architecture and shorter pods are appropriate objectives of selection. Dwarf cultivars perhaps do not produce sufficient leaf area for high yields.

Prospects Compared with common bean, runner bean is underexploited. It obviously could be used as a vegetable and as a pulse crop in highland tropical areas with a suitable rainfall regime. Increasing interest in vegetarian diets and the replacement of plant by animal protein might lead to increased interest in seed production. A promising species for South-East Asia too.

Literature **1** Duke, J.A., 1981. Handbook of legumes of world economic importance. Plenum Press, New York and London. pp. 189–191. **2** Westphal, E., 1974. Pulses in Ethiopia: their taxonomy and agricultural significance. Pudoc, Wageningen. Agricultural Research Reports 815. pp. 135–140.

J. Smartt

Phaseolus lunatus L.

Species Plantarum 2: 724 (1753).

LEGUMINOSAE

$2n = 22$

Synonyms *P. bipunctatus* Jacq. (1770), *P. limensis* Macfad. (1837).

Vernacular names Lima bean, butter bean, Madagascar bean (En). Haricot de Lima, pois du Cap, fève créole (Fr). Indonesia: Java bean, kratok. Malaysia: kacang China, kacang Jawa, kekara kratok. The Philippines: sibatse simaron, patáni, zabache. Burma: Burma bean, white Burma. Thailand: thua rachamat. Vietnam: dâu ngu.

Origin and geographic distribution Lima bean has a Neotropical origin with at least two centres of domestication: Central America (Mexico, Guatemala) for the small-seeded forms and South America (mainly Peru) for the large-seeded forms. In post-Columbian times, Lima bean spread throughout America. Spaniards took seeds across the Pacific to the Philippines and from there to Asia, mainly Java and Burma, and to Mauritius. The slave trade took them from Brazil to Africa, particularly in the western and central parts. Some large-seeded forms from the Peruvian coast were distributed to south-western Madagascar and southern California. It is now cultivated throughout the warmer parts of the world.

Uses The crop is cultivated primarily for its immature and dry seeds. Particularly in Asia, immature sprouts, leaves and pods are consumed too. In the Philippines, dry seeds are used to produce a bean flour rich in protein to enrich bread or noodles. Seeds and leaves are valued for their astringent qualities and consequently used as a diet for fever in traditional Asian medicine. After harvesting the pods, the vines are sometimes offered to cattle in Malaysia and Indonesia. Lima bean is also grown as a short-duration cover crop or for green manure in Malaysia.

Production and international trade Production statistics from many tropical regions are fragmentary and often aggregated with other pulses. The United States is the world's largest producer with 20 600 ha under cultivation and an annual production of green beans of 55 600 t in 1980. Madagascar is the second largest commercial producer with an area cropped of about 7000 ha and a production of dry seed about 8000 t, almost exclusively of the large white-seeded types. Peru came third with production of dry seed averaging 5000–5500 t from about 5000–6000 ha. In other countries,

Lima beans are grown mostly in gardens or as an intercrop in fields, without any accurate estimate of area or production. In Asia, Burma is the major producer but no figures are available.

Properties Dry seeds contain per 100 g edible portion: water 13.2 g, protein 14.4–26.4 g, fat 1.5 g, carbohydrates 58.0 g, fibre 3.7 g, ash 3.4 g. The energy content averages 1450 kJ/100 g. Main limiting amino acids are methionine and cystine (with respect to total N 1.1–1.2 g per 16 g). Per 100 g edible portion green beans contain 1.3 g protein, green seeds 8.4 g, green leaves 0.6 g. Antimetabolic factors include protease inhibitors, lectins and cyanogenic glucosides (linamarin). Seed weight varies between 30 and 300 g/100 seeds. In Nigeria, cooking time of dry seeds is 60–90 minutes.

Description An annual or occasionally perennial herb, bushy forms up to 0.6 m, climbing forms 2–4 m tall. Roots thin or swollen, up to 1.5–2 m deep. Leaves trifoliolate; leaflets ovate and acuminate, 5–19 cm × 3–11 cm. Inflorescences axillary racemes, up to 15 cm long, with many nodes and flowers; bracteoles persistent; calyx campanulate; corolla 0.7–1.0 cm wide; standard hood-shaped, pale green or violet; wings white or violet; keel sharply upturned, white or occasionally pigmented; stamens ten, diadelphous; style coiled with pubescent apical region; stigma ellipsoid, directed adaxially. Pods oblong, 5–12 cm × 2.5 cm, generally curved, sometimes with a hook-shaped top, with 2–4 seeds. Seeds variable in size, shape and colour; kidney-shaped, rhomboid or round; colour uniform or speckled or mottled, white, green, yellow, brown, red, black or purple; often transverse lines radiate out from the hilum. Seedlings with epigeal germination, first leaves simple and opposite.

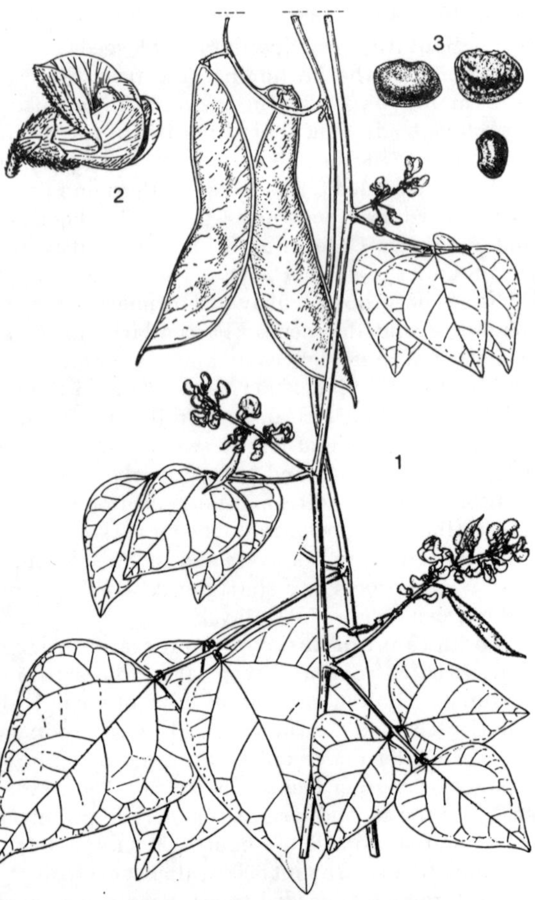

Phaseolus lunatus L. – 1, flowering and fruiting branches; 2, flower; 3, seeds.

Growth and development Germination and emergence occur 4 to 10 days after sowing. Vegetative growth accelerates after 1 month. Flowers appear 35–70 days and ripe pods 80–120 days after sowing with short daylength. In climbing types, flowering and fruiting may extend throughout the wet season. Pollen and stigma mature synchronously and in close proximity within the unopened bud, favouring self-pollination. However cross-pollination often occurs too. Of buds, flowers and young pods, 75–85 % are shed.

Other botanical information Within the species, there are wild plants and cultivated plants. Both groups are distinguished as botanical varieties: var. *silvester* Baudet, var. *lunatus*, respectively. More in accordance with the International Code of Nomenclature of Cultivated Plants is the distinction of cultivar (cv.) groups below species level for the cultivated plants: cv. group Sieva with medium-sized flat seeds, cv. group Potato with small globular seeds and cv. group Big Lima with large flat seeds.

Cultivars grown in Asia are usually classified into 4 groups.

- Java beans: medium-sized, purplish-red to black with large amounts of HCN.
- Red Rangoon or Red Burma beans: small reddish beans that are usually plump and sometimes with purplish spots. They contain traces of HCN.
- White Rangoon or White Burma beans: small white beans, usually plump and resembling small haricots. They usually contain traces of HCN.
- Lima beans: large plump white beans. Said not to contain HCN.

Ecology Lima bean contains day-neutral genotypes that flower in daylengths of 9–18 hours and short-day types that require critical daylength of 11–12.3 hours to initiate flowers. Optimum temperatures are 16–27 °C; frost is not tolerated. Normal rainfall is 900–1500 mm/year but the crop tolerates as little as 500–600 mm, once established. It is grown in lowland tropical and subtropical areas but may reach 2000–2500 m. Lima bean prefers well aerated well drained soils with pH 6.0–6.8. However some cultivars tolerate acid soils with pH as low as 4.4.

Propagation Propagation is by seed. In the tropics, Lima bean is cultivated in home gardens or intercropped with cereals (maize, sorghum), roots and tubers (yam, cassava) or other crops (e.g. cotton, sugar-cane). Sole cropping is more frequent in the United States, Madagascar and Peru.

Husbandry Bush types are usually spaced 20–30 cm within rows and 60–100 cm between rows, while climbing types may be planted on hills 90–200 cm apart. Seeds are often placed in the same hill as the companion crop in intercropping. Weeding is necessary during initial growth. In humid areas, climbing types are staked; in drier conditions (California, Madagascar and Peru), they may be left prostrate and irrigated 2–4 times before maturity. Fertilizer is not commonly used in tropical areas. In the United States, Madagascar and Peru, Lima bean is often grown as a sole crop; in other regions, rotation with food or cash crops is more frequent.

Diseases and pests In the tropics, the most serious diseases are web blight caused by *Rhizoctonia* sp., anthracnosis caused by *Colletotrichum* sp. and golden mosaic caused by a virus transmitted by white flies (*Bemisia* sp.). Root-knot nematodes (principally *Meloidogyne*), pod borers (*Maruca* sp., *Cydia* sp. and *Etiella* sp.) and bruchids (*Acanthoscelides* sp., *Zabrotes* sp.) are considered serious pests. The crop is, however, rugged in traditional cropping systems.

Harvesting Green and mature pods of the climbing types are usually picked manually over a long period of time. In drier areas (Madagascar), whole plants are cut and left to dry in the field before the pods are removed. Mechanical picking is possible with erect cultivars maturing uniformly and setting pods well above the soil surface.

Yield Throughout the tropics, yield of dry seed is 200–600 kg/ha in intercropping and 1000–1500 kg/ha in sole cropping. In trials, production of dry seed in pure stands may reach 2000–2500 kg/ha for the bush types and 3000–4000 kg/ha for the climbing types.

Handling after harvest Pods, sometimes whole plants, are threshed usually by hand, and seed is cleaned and sorted. Care should be taken with threshing, as the seeds are brittle and easily damaged. In many tropical countries, seeds are sometimes stored in jars or baskets, and covered with a layer of sand or ash to protect them against bruchid infestation. In the United States, green Lima beans are canned or frozen to eat as a vegetable.

Genetic resources Over 2500 samples are available in the CIAT collection at Cali (Colombia) with seeds coming mainly from South and Central America, West Africa, Madagascar and Burma. Other smaller gene banks are in the United States, Brazil, Costa Rica and Nigeria (IITA, Ibadan). More than half of the primary gene pool is still uncollected and surveys are urgently needed in the primary centres of diversity of South and Central America, but also in the secondary centres of West Africa, Madagascar and South-East Asia.

Breeding Erectness, resistance to lodging and to web blight are prime criteria to improve bush types. Earliness, photoinsensitivity, resistance to golden mosaic and suitability for intercropping are being sought in climbing types. Some promising genotypes in the humid tropics have been identified among the climbing forms. A large secondary gene pool is available for improvement and the following wild species have been successfully crossed with Lima bean: *P. jaliscanus* Piper, *P. metcalfei* Wooton & Standley, *P. polystachyus* (L.) Britton, Sterns & Poggenb., *P. ritensis* M.E. Jones, *P. salicifolius* Piper and two other new yet unnamed species. Introgression of useful genes from the wild taxa (e.g. resistance to golden mosaic) has been observed in interspecific breeding material.

Prospects Lima bean offers good prospects in both the subtropicals and the semiarid to humid tropics. Deep rooting, drought tolerance and high-yielding potential are useful attributes for wide adaptation. Research priorities should first be devoted to full exploitation of the large genetic variation available in the primary gene pool.

Literature |1| Baudoin, J.P., 1982. Essais d'évaluation du haricot de Lima (Phaseolus lunatus L.) en vue de l'amélioration de sa culture en régions tropicales de basse altitude. Bulletin de Recherche Agronomique, Gembloux 17(1):3–16. **2** Duke, J.A., 1981. Handbook of legumes of world economic importance. Plenum Press, New York and London. pp. 191–195. |3| Kay, D.E., 1979. Food legumes. Tropical Products Institute, London. Crop and Product Digest No 3. pp. 225–245. |4| Lyman, J.M., Baudoin, J.P. & Hidalgo, R., 1985. Lima bean (Phaseolus lunatus L.). In: Summerfield, R.J. & Roberts, E.H. (Editors): Grain legume crops. Collins, London. pp. 477–519.

J.P. Baudoin

Phaseolus vulgaris L.

Species Plantarum 2: 723 (1753).
LEGUMINOSAE
$2n = 22$

Synonyms *Phaseolus esculentus* Salisb. (1796).

Vernacular names Common bean (En). Haricot commun (Fr). Indonesia: buncis. Malaysia: kacang buncis, kacang merah. The Philippines: butingi, mula. Burma: bo-sa-pè, pè-bya-galè, pè-gya(ni). Cambodia: sândaèk barang. Laos: thwàx fàlangx. Thailand: thua khaek, thua phum. Vietnam: dâu ve.

Origin and geographic distribution Common bean originated in the Americas and domesticated in Mexico, Peru and Colombia 8000 years ago. Before Columbus, it was unknown in the Old World, but later it has become a major crop in Europe and Africa. It is widely cultivated in many parts of the tropics and subtropics, and throughout the temperate regions. Common bean is the main pulse crop throughout tropical America and many parts of tropical Africa. It is a minor crop in India and most of tropical Asia, where indigenous pulses are preferred.

Uses Common bean is used as a pulse and as a green vegetable. The young pods and ripe seeds are eaten and, to a lesser extent, also the green-shelled seeds. In some parts of the tropics, young leaves are used as a spinach. In temperate regions, common bean is grown mainly for the green immature pods that are eaten as a vegetable, and are also canned and frozen. The dried seeds are also cooked with tomato sauce and canned. The straw is used as fodder. Preparation is by boiling; beans are very acceptable with a wide range of meat and vegetable sauces. The straw can be used as forage.

As a pulse, common bean can be used interchangeably with cowpea. Their use, however, is complementary since temperature optima for the cowpea are higher than for the common bean.

Production and international trade Production data for 'dry beans' in FAO statistics must be treated with caution. That category includes products from species of *Vigna* and other genera. Production of common bean alone accounts for 95 % of the total world production of *Phaseolus* bean, which is 8,3 million t. Less than 5 % of total production is from the other three cultivated species (estimates from world production of dry beans in 1979, subtracting known non-*Phaseolus* species). Common beans are grown extensively in five major continental areas: eastern Africa, North, Central and South America, estern Asia, and western and south-eastern Europe. Roughly 30 % of world production is in tropical America. Brazil is by far the largest bean-producing nation in the world with 2.3 million t (including some cowpeas) over the period 1973–1975.

Properties Per 100 g edible portion dry seeds contain: water 10 g, protein 22.6 g, fat 1.4 g, carbohydrates 62 g, fibre 4.3 g and ash 3.7 g. The energy content averages 1453 kJ/100 g. For green pods, contents are: water 91 g, protein 1.8 g, fat 0.2 g, carbohydrates 6.6 g, fibre 1 g and ash 0.7 g/100 g. The energy content averages 126 kJ/100 g. For fresh leaves, protein 3.6 g/100 g, a high vitamin C content (110 μg/100 g), high contents of vitamin A precursors and energy content 151 kJ/100 g. The contents vary with genotype and production locality. Dry beans must be thoroughly cooked before being eaten, since raw beans contain an array of antimetabolites. Beans contain significant amounts of thiamine, niacin, folic acid and fibre. In general, deep red and black beans are richer in tannins than pale and lightly variegated types, and tannins are considered to reduce the dietary value of the protein. Weight varies between 20 to 200 g/100 seeds.

Description A climbing, viny or bushy, slightly pubescent annual with well developed tap-root, laterals and adventitious roots. Sometimes globular nodules are present. Stem angular or nearly cylindrical. Leaves alternate, trifoliolate; petiole usually up to 15 cm long, grooved above, with basal pulvinus; basal leaflets asymmetrical; apical leaflet symmetrical, usually ovate, 7.5–14 cm × 5.5–10 cm. Inflorescence axillary or terminal racemes with several to many white, pink, lilac or purple flowers; calyx campanulate; corolla papilionaceous; stamens diadelphous (9 + 1); ovary laterally compressed with 4–12 (commonly 7) ovules; style upturned and spiralled with collar of fine hairs below the stigma; stigma ellipsoid, glandular, directed adaxially. Pod linear, up to 20 cm long, straight or more commonly curved with a prominent beak, fleshy when immature, green or yellow, sometimes purple or reddish to purplish spotted or striped. Seeds highly variable in size, shape and colour; ovoid, subspherical or kidney-shaped; black, brown, yellow, red and white with

Phaseolus vulgaris L. – 1, fruiting branch; 2, inflorescence; 3, seeds.

speckled, flecked, saddled and darker patterns superimposed. Seedling with epigeal germination; first two leaves simple and opposite, subsequent leaves alternate, trifoliolate.

Growth and development Seedlings usually emerge within a week after sowing at a soil temperature of 16 °C; at lower temperatures, they may take up to 2 weeks. Two growth types are distinguished: determinate, in which the main axis terminates in an inflorescence and produces no vegetative nodes after flowering, and indeterminate. Determinate plants of common bean have a central axis (the main stem) with 5–9 nodes and from two to several branches that arise from the more basal nodes. Indeterminate plants have a central axis with 12–15 nodes or even more in climbing viny types. Time to flowering varies with cultivar, temperature and photoperiod, and is usually 28–42 days. Flowering is usually completed in 5–6 days at 20–25 °C in determinate bush genotypes and in 15–30 days in indeterminate climbing genotypes. Climbing cultivars can flower over a much longer period, if immature pods are harvested. Flowers open at sunrise and fade at sunset. Self-pollination is usual; frequency of cross-pollination is low. As many as two thirds of the flowers produced may abort and, under temperature or moisture stress, young fruits and developing seeds may abscise as well. Abscission is most frequent in flowers formed on the later nodes and branches, and in the later flowers or racemes with multiple flowers. Seed-filling period may take from as few as 23 days to nearly 50 days. Full maturity with dry seed is reached 65–150 days after sowing.

Other botanical information CIAT has classified the world collection into four main types by growth habit on the basis of determinacy, node production after flowering, height and climbing tendency:

- dwarf determinate habit, with reproductive terminals on main stem and no further node production on main stem after flowering (short, self-supporting or bushy type of short growth duration);
- dwarf indeterminate habit, vegetative terminal on main stem and further node production on main stem after flowering, with erect branches borne on lower nodes;
- prostrate indeterminate habit, moderate to considerable node production on main stem after flowering, with variable number of branches borne on lower nodes and with prostrate to cone-shaped canopy on supports;
- climbing indeterminate habit, heavy production of nodes on main stem after flowering, with branches not well developed compared to main stem development, with moderate to strong climbing-ability on supports.

Pod texture is closely related to use. Leathery pods, with reduced parchment tissue are used to produce both dry seed and green beans as they remain tender until the rapid phase of seed growth occurs. Stringless pods remain tender when seeds are large and are extensively traded frozen.

Generally beans are classified according to their use as follows:

- snap or string beans, grown for the pods harvested before being fully grown and, while still slender, with small seeds (e.g. green-podded bush cultivars, wax or yellow-podded bush cultivars, green-podded climbing cultivars, wax or yellow-podded climbing cultivars);
- green-shell beans, used in the green-shelled condition (bush cultivars, climbing cultivars);
- dry-shell or field beans, grown for the dry, ripe seeds (medium field beans with seeds 1–1.2 cm long, pea or navy beans with seeds up to 8 mm long, red kidney beans with seeds 1.5 cm long or more, and marrow beans with seeds 1–1.5 cm long).

Ecology Common bean is a quantitative short-day plant. For every genotype, there is an optimum for photoperiod and temperature where that genotype will flower after the smallest possible time interval from emergence. Deviations in either temperature or photoperiod cause delays in flowering. Most common beans are grown within a narrow range of temperatures (17.5–)20–22.5(–25) °C; in the equatorial tropics, they are found at higher altitudes (above 1000 m). Temperatures below or above the optimum reduce yield through plant mortality (at high temperatures), reduced photosynthesis and failure of flowers to produce mature pods (50–70 % of opened flowers). Common beans are sensitive to night frost. A moderate well distributed rainfall is required (300–400 mm per crop cycle) but dry weather during harvest is essential. Drought or waterlogging are detrimental. Suitable soil types range from light to moderately heavy and to peaty soils with near-neutral pH and good drainage. Common bean is susceptible to salinity.

Propagation Normal propagation is by seed but for special purposes stem cuttings can be rooted easily. Plant population densities are 150 000–200 000 plants/ha for dwarf cultivars and half that for climbing types in sole cropping. In intercropping, densities are much lower. Plant distance varies for bush cultivars in sole cropping 30–45 cm between the rows and 30 cm in the rows. Wider spacing to 75–90 cm × 10–15 cm makes weeding easier. For climbing beans, 4–6 seeds are usually sown together in hills spaced 1 m apart. They may also be sown in rows at a spacing of 90–120 cm × 15–30 cm. Depth of sowing is 3–6 cm. Seed rate depends on seed size and intended plant population densities, up to 120 kg/ha for dwarf beans and 60 kg/ha for climbers in sole cropping.

In North America, Europe and in limited areas of other producing regions, common beans are extremely commercialized. Bush types predominate there, since they are well suited to intensive cultivation. Most sole-cropped beans are grown as a high-input crop with yield potentials 1000–3000 kg/ha. In subsistence cropping, landraces of different plant types and seed colours are grown in complex intercropping systems. Of the common beans in tropical America, 75–80 % are in some form of intercropping, mostly with maize.

Husbandry Flat cultivation is preferred to ridges. Disturbance of the soil should be avoided because damage to the roots or the collar of the plant encourages diseases. So shallow tillage is preferred especially in the period before flowering. Common bean can be rain-fed or irrigated. Irrigation is beneficial in semiarid regions, with overhead irrigation preferred over flood irrigation. In peasant farming, the crop is seldom manured. Crop rotation is necessary to limit disease.

Diseases and pests Common diseases in the tropics are web blight (*Rhizoctonia microsclerotia*), common blight (*Xanthomonas phaseoli*), root-rots (*Fusarium* spp. *Rhizoctonia* spp. and *Macrophomina* spp.), rust (*Uromyces phaseoli*), anthracnose (*Colletotrichum lindemuthianum*), angular leaf-spot (*Isariopsis griseola*), the viruses common mosaic (BCMV) and golden bean mosaic (GBMV).

Control of seed-borne diseases is achieved by sowing disease-free seed, good crop rotation and disposal of infected plant residues. Genetic resistance to rust, anthracnose and angular leaf-spot is useful but production of new virulent biotypes is a problem.

Soil-borne diseases and nematode pests (*Meloidogyne* and *Pratylenchus*) are difficult to control except by crop rotation, though potentially useful genetic resistance is known.

Insect pests, though numerous, especially in tropical America, are often only of local significance. Many species of aphids and leafhoppers are pests of common beans, as well as numerous lepidopterous caterpillars and chrysomelid beetles. Beans beetle (*Acanthoscelides obtectus*) and cowpea beetles (*Callosobruchus* spp.) are widespread in the tropics, attacking filling and dried beans. The most serious pests of the dried pulse are bean weevils (*Bruchus* spp.). The beanfly (*Melanagromyza phaseoli*) is a serious pest in Africa and Asia.

Control of pests by chemicals such as carbaryl can be useful. Complete control of bruchids can be achieved by coating stored seeds with cotton seed oil. Pyrethrins are effective in controlling weevils. Genetic resistance might be incorporated in new cultivars.

Harvesting Snap beans are harvested before the pods are fully grown. Harvest starts 7–8 weeks after sowing in early cultivars. Pods should be picked every 3–4 days and the number of pickings is greater in climbing cultivars than in bushy ones. Dry beans are harvested as soon as a considerable proportion of the pods are fully mature and have turned yellow. Some cultivars tend to shatter. Usually entire plants are pulled. Large-scale crops are harvested mechanically.

Yield Yield of seed may be in excess of 1.3 t/ha but may be half of that or less under poor cultural conditions. Yield of green pods may reach 5 t/ha.

Handling after harvest Harvested plants are

dried for a few days in the windrow, stacked and threshed when dry. Seed should be dry, clean and free from mechanical damage before being stored under optimum conditions. Dry beans of all types may be canned successfully, whereas immature pods of certain cultivars of snap bean are a popular processed vegetable.

Genetic resources The primary gene pool is well represented in collections of CIAT, the United States and Soviet Union and many smaller national collections. The wild component of that gene pool requires further exploration and collection between Mexico and Argentina. The secondary gene pool consists of *P. coccineus* L. and *P. polyanthus* Greenman. The tertiary gene pool includes most other *Phaseolus* species.

Breeding Breeding objectives include resistance to the common pests and diseases, production of lectin-free genotypes, improved nitrogen fixation capacity especially in dwarf early-maturing cultivars, improved canopy architecture in relation to improved yielding capacity and resistance or tolerance to environmental stress.

Prospects Factors against the common bean are the long cooking they require, antinutritional factors, a flatulence factor and susceptibility to the development of hard-shell and off-flavours during storage. Improvement of storage and processing methods, and development of new food products could increase consumption considerably in many producing countries. Increased efficiency of production will require improved morphological, ecological, physiological and biochemical efficiency of the plant, for which the available genetic resources are adequate.

Literature |1| Adams, M.W., Coyne, D.P., Davis, J.H.C., Graham, P.H. & Francis, C.A., 1985. Common bean (Phaseolus vulgaris L.). In: Summerfield, R.J. & Roberts, E.H. (Editors): Grain legume crops. Collins, London. pp. 433–476. |2| CIAT, 1976–1980. Abstracts on field beans (Phaseolus vulgaris L.). Vol. 1–5. Cali, Colombia. |3| Duke, J.A., 1981. Handbook of legumes of world economic importance. Plenum Press, New York and London. pp. 195–200. |4| Janssen, W., 1988. Snap beans: present status in the developing world and bibliography of research (1919–1987). CIAT, Cali, Colombia. 411 pp. |5| Kay, D.E., 1979. Food legumes. Tropical Products Institute, London. Crop and Product Digest No 3. pp. 124–176. |6| López, M., Fernández, F. & van Schoonhoven, A. (Editors), 1985. Frijol: investigación y producción. PNUD & CIAT. Cali, Columbia. 417 pp. (English edition in preparation).

J. Smartt

Pisum sativum L.

Species Plantarum 2:727 (1753).
LEGUMINOSAE
$2n = 14$

Vernacular names Pea (En). Pois (Fr). Indonesia: kapri, kacang polong. Malaysia: kacang ercis. The Philippines: citzaro. Burma: sadaw-pè. Cambodia: sândaèk muul. Vietnam: dâu hoà lan.

Origin and geographic distribution Probably originated in South-West Asia; now cultivated in many temperate countries, as a cool-season crop in the subtropics and at higher altitudes in the tropics.

Uses Dry seeds are used for food and feed. For the former, they are cooked whole, split or ground into flour, and boiled or roasted. Large amounts are canned. Fresh peas are a major vegetable and they are canned or frozen, in the immature form. Some cultivars are grown for their tender green pods. The species is suitable as forage, hay, silage and green manure too.

Production and international trade World production for dry and green peas in 1985 was 11.6 million t and 4.9 million t from areas of 8.9 million ha and 778 000 ha, respectively. The world price for dry peas in 1987 was US$ 480 per tonne. In the Far East, 520 000 t were produced in 1986 from 675 000 ha.

Properties Mature round seeded cultivars have per 100 g edible portion: water 10–13 g, protein 16–33 g, fat 1–2.5 g, carbohydrates 49–62 g (starch 43-56 g, sugars 6 g), fibre 4.5 g, ash 2–3 g. The energy content averages 1428 kJ/100 g. Wrinkled forms have starch 27–37, sugars 10, similar amounts of protein, though with less sulphur amino acids than round forms, and fat 5 g/100 g. Peas are relatively free from antinutritional factors. Cultivars with a dark testa contain tannin which can lessen their digestibility. Weight of 100 seeds varies from 3 to 48 g.

Botany An annual climbing herb, primitive forms up to 3.5 m tall, modern cultivars as short as 30 cm. Leaves pinnate with 2–3 pairs of leaflets and a terminal tendril, at base with 2 leaf-like stipules; leaflets sometimes converted to tendrils. Flowers in axillary racemes with 1–2 flowers, calyx of 5 green united sepals and corolla of 5 white, purple or pink petals; anthers 10, diadelphous; ovary with 4 to 15 ovules. Fruit 3–11 cm long, sometimes with a tough inner membrane, with 3–11 seeds. Seeds round or wrinkled, varying in colour. They germinate rapidly (hypogeal) and develop to maturity in 80–150 days. Flower initia-

Pisum sativum L. – 1, branch with flower; 2, branch with fruit; 3, seed.

tion varies markedly with cultivar and daylength but in long days can be within three weeks of sowing at 18 °C. The period of flowering can be as little as 2–3 weeks in modern cultivars but is much longer in more primitive forms. The flowers are strictly self-pollinated. Classification is mainly by seed type, whether round or wrinkled; the former are grown mainly for harvesting as mature seed, and the latter for canning and freezing in the immature form.

Ecology Peas grow reasonably well between 10 and 30 °C with an optimum of 20 °C. A rainfall as low as 400 mm/year will suffice, with an optimum of 1000 mm/year. In tropical regions, the crop has to be grown above an altitude of 1200 m. Peas tolerate a wide range of soils but thrive best on a well drained one that is slightly acid. Time of sowing and place in the crop rotation depend on regional climate. Dry peas are primarily grown as a break or catch crop in cereal rotations.

Agronomy Seed is sown at a depth of 4–7 cm at a rate of 23 g/m^2 or 65–280 kg/ha, with the higher rates for garden peas. Seed-bed compaction and weed competition should be avoided. Pea weevil (*Sitona lineatus*) is a common pest but is controlled by applying insecticides during flowering. Root-rots caused by a complex of fungal pathogens are avoided by adequate crop rotation lasting at least 3–5 years. In humid areas, downy mildew (*Peronospora pisi*) and, in hotter areas, powdery mildew (*Erysiphe pisi*) are serious fungal pathogens. Virus diseases can be a problem. Average yields in Asia and the world are 1.2 t/ha and, in France, 4 t/ha. For green peas, the yields for those three respective areas are 4, 6 and 7.5 t/ha.

Genetic resources and breeding There are collections of germplasm in 24 centres. The wild relatives are primitive forms. Many useful genes for disease resistance are found in the centres of diversity in the Near and Middle East. The main breeding objectives are improving the standing capacity of the crop, disease resistance and higher yields. Selection for tolerance to stress factors such as drought, temperature extremes and waterlogging should improve the stability of yields.

Prospects Peas are in heavy demand for culinary purposes in the tropics too. Improved yields through better standing capacity and disease resistance should readily be obtained.

Literature |1| Ayad, G. & Anishetty, N.M., 1980. Directory of germplasm collections. 1. Food legumes. IBPGR, Rome. |2| Davies, D.R., Berry, G.J., Heath, M.C. & Dawkins, T.C.K., 1985. Pea (Pisum sativum L.). In: Summerfield, R.J. & Roberts, E.H. (Editors): Grain legume crops. Collins, London. p. 147–198.

D.R. Davies

Vicia faba L.

Species Plantarum 2: 737 (1753).
LEGUMINOSAE
$2n = 12$

Vernacular names Faba bean (En). Fève (Fr). Indonesia: kacang babi. Burma: sandusi. Thailand: thua yang.

Origin and geographic distribution The origin of the faba bean is in the Mediterranean Region or in South-West Asia, where it has been cultivated since ancient times. It is widely cultivated in all temperate regions and at higher altitudes in the tropics. It ranks among the world's main grain legumes.

Uses Faba bean is grown as garden cultivars for the green shell beans and as a field crop for the dried beans that are used as food or feed. They are also grown for fodder, hay and green manure (whole plants or harvested plants).

Production and international trade In 1981, faba bean was grown in more than 50 countries, on 3.6 million ha with a production of beans of more than 4 million t. The largest producer is China (60 % of total area, 65 % of world production), second is Ethiopia with 9 % of the world area.
From South-East Asia, production is reported from Indonesia, Thailand and Burma, without data.

Properties Per 100 g edible portion, whole dried seeds contain: water 10 g, protein 26 g, fat 1 g, carbohydrates 59 g, fibre 7 g, ash 3 g. The energy content averages 1450 kJ/100 g. Seed weight varies between 40 and 180 g/100 seeds.

Botany An erect, stiff, glabrous annual, 30–180 cm tall. Stem stout, square and hollow with 1–7 branches at the base. Leaves alternate, pinnately compound; leaflets 2–6, ovate to elliptic, 3–10 cm × 1–4 cm. Inflorescences axillary short racemes with 1–6 flowers, which are 2–4 cm × 1.5 cm, white with black or purplish streaks and blotches, fragrant. Pods subcylindrical to flattened, in field cultivars 5–10 cm long, in garden cultivars up to 30 cm long. Seeds very variable in shape and size, strongly compressed to nearly spherical, 1–2.6 cm long, white, green, buff, brown, purple or black.
Germination is hypogeal. Cropping season duration varies widely from 3 months (Sudan, Canada) to 11 months (e.g. winter beans in north-western Europe). In African and Indian cultivars, supra-optimum temperatures at flowering time can adversely affect fertilization.
Faba beans are mainly cross-pollinated by insects. Flowering progresses from the lower part to the top part of the stem and takes 14–20 days.
Four cv. groups are distinguished.
- cv. group Major: the broad bean, with large flattened seeds, average length 2.5 cm;
- cv. group Equina: the horse bean, with medium-sized seeds, average length 1.5 cm;
- cv. group Minor: the tick bean, with small rounded seeds, about 1 cm long;
- cv. group Paucijuga: with less than 4 leaflets per leaf.

Vicia faba L. – 1, flowering and fruiting branch; 2, seeds.

Ecology Faba bean is an annual from temperate regions. It can be grown at high altitudes in the tropics and in winter in the subtropics, but it usually fails to produce pods in the hot humid tropics. It is a quantitative long-day plant; some cultivars are day-neutral. During growth, an average temperature of 18–27 °C is required, with little or no excessive heat, some cultivars tolerate frost. Rainfall needs to be 650–1000 mm/year; moisture requirement is highest 9–12 weeks after establishment. It is not drought-resistant and cannot tolerate waterlogging. It tolerates nearly any soil type but grows best on rich loams; pH may be 4.5–8.3, optimum 6.5.

Agronomy Faba bean is propagated by seed. Seeds are usually sown 5–10 cm deep in rows 75 cm apart, within the rows 15 cm apart. Recommended rate of sowing for small-seeded cultivars is 90–122 kg/ha, for large-seeded cultivars 78–90 kg/ha, but up to 450 kg/ha is used. For green manure or forage, small-seeded cultivars are usually broadcast. Faba beans are cultivated by smallholders and by estates.
Faba beans should be carefully tended throughout their growing period. Post-emergence and pre-emergence herbicides are often applied. In Egypt and Spain, faba bean is often irrigated.
Fertilizers and seed inoculation with proper legume bacteria are recommended. In temperate regions, faba beans are usually alternated with wheat or barley, in warm climates with rice, cot-

ton, maize, sorghum, rapeseed and sweet potato. In tropical areas, the application of superphosphate (phosphoric acid) and potash, at rates of 50–100 kg/ha and 25–50 kg/ha, respectively, are recommended.

For animal fodder, faba beans are sometimes grown intercropped with cereals, peas or *Lathyrus* spp.

Serious diseases are caused by *Botrytis fabae*, *Ascochyta*, *Uromyces* and *Fusarium*. Virus, nematode and bacteria diseases are numerous. The most serious insect pests are faba bean weevil, *Bruchus rufimanus* and bean aphid, *Aphis fabae*. In the Middle East, broomrape *Orobanche crenata* may be a serious problem.

Faba beans mature 90–220 days after planting. The crop should not be cut until the lower pods have matured and the upper ones fully developed. The crop should preferably be cut on a cloudy day or at night, and gathered early the next day. Mechanical harvesting is possible. If grown for forage, faba beans are normally harvested shortly before the end of flowering when the first pods are well formed.

Yield of dry beans varies with region and cultivar. In 1981, yield in Great Britain was 2700 kg/ha, in China 1230 kg/ha, in Egypt 2500 kg/ha, in Ethiopia 850 kg/ha and in Mexico 1720 kg/ha; world average was 1154 kg/ha.

Yield of fresh green faba beans to eat at home as a vegetable averages 11–12.5 t/ha in Great Britain. After threshing, seeds are cleaned with ordinary fanning mills. For canning, beans are allowed to fill on the plant and are picked by hand before they become hard. As a dried vegetable, they are prepared in the same way as other common beans.

Genetic resources and breeding There are collections of faba bean germplasm at ICARDA (Syria), the University of Bari (Italy), at Brunswick (West Germany) and at Gatersleben (East Germany).

The main breeding objectives are complete self-fertilization, determinate growth habit, shorter racemes, improved harvest index, white flowers and testas free from tannin.

V. faba seems to be completely isolated genetically from the other 148 known *Vicia* species. Within *V. faba*, variability is great and more collections are needed from the centres of origin.

Prospects There is renewed interest throughout the world for the faba bean, because costs of protein-rich food and feed are rising, because of the need to diversify and because of increasing human populations. So faba beans could well have a promising future. Through international cooperation, improved cultivars are appearing. In South-East Asia, faba bean is an interesting crop for high mountainous regions.

Literature |1| Bond, D.A., Lawes, D.A., Hawtin, G.C., Saxena, M.C. & Stephens, J.H., 1985. Faba bean (Vicia faba L.). In: Summerfield, R.J. & Roberts, E.H. (Editors): Grain legume crops. Collins, London. pp. 199–265. |2| Duke, J.A., 1981. Handbook of legumes of world economic importance. Plenum Press, New York and London. pp. 275–279.

P.C.M. Jansen

Vigna aconitifolia (Jacq.) Maréchal

Bulletin du Jardin Botanique National de Belgique 39: 160 (1969).

LEGUMINOSAE

$2n = 22$

Synonyms *Phaseolus aconitifolius* Jacq. (1768).

Vernacular names Moth bean (En). Haricot papillon (Fr). Malaysia: mittikelu. Thailand: matpe.

Origin and geographic distribution Moth bean is a native of India, Pakistan and Burma, where it grows both wild and cultivated. It is also grown in China, Sri Lanka, Africa, United States and Thailand.

Uses Green pods and ripe whole or split seeds are cooked and eaten. In India and the United States, moth bean is also grown for green manure, forage and hay and as cover crop. Seeds are used medicinally in diets for fevers; roots are said to be narcotic.

Production and international trade Recent data are not available. It is estimated that 1.8 million ha are under this crop. In 1974, India was the largest producer of moth beans with 500 000 t/year. Thailand is the leading exporter with 30 000–40 000 t/year.

Properties Per 100 g edible portion mature seeds contain: water 10.8 g, protein 23.6 g, fat 1.1 g, carbohydrates 56.5 g, fibre 4.5 g, ash 3.5 g. Seed weight is 1 g/100 seeds.

Botany Annual herb with short erect stem, 10–40 cm long, with many prostrate branches 30–150 cm long. Leaves alternate, trifoliolate; leaflets up to 5-lobed, 5–12 cm long, lobes narrow, acuminate. Flowers about 9 mm long, yellow, several together in axillary capitate racemes with peduncles 5–10 cm long. Pods subcylindrical, 2.5–5 cm × 0.5 cm, covered with short brown hairs, contain-

Vigna aconitifolia (Jacq.) Maréchal – 1, flowering branch; 2, fruiting branch; 3, seeds.

ing 4–9 seeds, which are rectangular to cylindrical, 3–5 mm × 1.5–2.5 mm, light brown, whitish green or yellow-brown, rarely mottled. Germination is epigeal. Vegetative development starts slowly. Faba bean takes 90 days to mature from sowing; early types mature in 75 days.

Ecology In India, moth bean is grown from sea level to an altitude of 1300 m. It requires uniform high temperatures and grows best with a soil temperature of 27 °C. It is drought-resistant and does well with a well distributed rainfall of 500–750 mm/year. It grows on many soil types but is particularly suitable for dry light sandy soils, and is an important crop of arid to semiarid regions in India. It is a quantitative short-day plant.

Agronomy Propagation is by seed, which should be sown on a well prepared seed-bed. Moth bean is usually sown broadcast, requiring 10–15 kg/ha when grown as a sole crop. When grown for forage, seed rate is 7–34 kg/ha. Sown in rows 75–90 cm apart at a depth of 2.5–4 cm, seed rate is 3.5–4.5 kg/ha for pure stands. In areas like the Indian arid belt, moth bean is interplanted with cereals, such as pearl millet and sorghum, or rotated as a green manure with cotton. It is irrigated or mostly rainfed.

Moth bean is notable for its pest and disease resistance, but it is susceptible to yellow mosaic virus, nematodes and leaf-spot diseases. It is also parasitized by *Striga* spp. Bean weevil attacks the seed during storage. Plants can be harvested 90 days after sowing. They are difficult to harvest with a mower because of the prostrate branches. The plants are usually cut with a sickle, left to dry for one week, then threshed and winnowed. Yields of seeds are 200–1500 kg/ha in India and 1240–1800 kg/ha in United States. Yield of green matter for forage is 37–50 t/ha and of hay 7.5–10 t/ha.

Genetic resources and breeding Low production is due to a lack of high-yielding cultivars. Germplasm collection is urgently needed. The ecological limits and the optimum methods of cultivation should be investigated. Breeding should concentrate on production of completely erect cultivars.

Prospects Lack of information on its potential and on its cultivation limits spread and use of moth bean. Because of its drought resistance, it could substantially increase production of food and forage in (semi)arid conditions, and protect the soil against erosion when other vegetation dies.

Literature |1| Duke, J.A., 1981. Handbook of legumes of world economic importance. Plenum Press, New York and London. pp. 286–288. |2| Mathur, J.R. & Sharma, R.C., 1980. Problems and prospects of moth bean (Vigna aconitifolia (Jacq.) Maréchal) in arid and semiarid zones of Rajasthan. In: Arid zone research and development. Scientific Publishers. Jodhpur, India. 531 pp.

C.C.C.M. van Oers

Vigna angularis (Willd.) Ohwi & Ohashi

Japanese Journal of Botany 44: 29 (1969).

LEGUMINOSAE

$2n = 22$

Synonyms *Phaseolus angularis* (Willd.) W.F. Wight (1909).

Vernacular names Adzuki bean (En). Haricot adzuki (Fr).

Origin and geographic distribution Adzuki bean is probably a native of Japan but has long been established in China and Sarawak. It is not known in the wild. It has long been cultivated in Japan, Korea, Manchuria, India and neighbouring areas of South Asia, and has been introduced to Hawaii, the southern United States, Angola,

Zaire, Kenya, Thailand, New Zealand and South America.

Uses The dried pulse is used as a human food, either cooked whole or made into a meal used in soups, cakes or confections. Sprouted beans are used as a vegetable. Beans may be popped like corn, used as a coffee substitute or eaten candied. The crop is grown also for forage and green manure. Flour is also used for shampoos and facial cream. In China, the beans are used to treat diseases like kidney trouble, constipation, abcesses, certain tumours, threatened miscarriage, retained placenta and non-secretion of milk. The leaves are said to alleviate fever and the sprouts are used to avert threatened abortion caused by injury.

Production and international trade No data for the global production of adzuki bean are available. Adzuki bean and adzuki bean flour are both large items of trade on oriental markets. China, Korea, Colombia and Thailand are the main exporters. Japan is the main producer (in 1970–1974 average 123 000 t/year) and consumer (1970–1974 average 27 875 t/year); it is second to soya bean there and commands highest prices.

Properties Per 100 g edible portion dry seeds contain: water 10.8 g, protein 19.9 g, fat 0.6 g, carbohydrates 64.4 g, fibre 7.8 g, ash 4.3 g. The energy content averages 1411 kJ/100 g. Seed weight varies between 10 and 20 g/100 seeds.

Botany An annual herb, usually bushy and erect, 25–90 cm tall, sometimes slightly vining or prostrate. Roots with many nodules. Leaves trifoliolate; leaflets ovate, usually entire, 5–9 cm long. Inflorescence axillary, short, with 6–12 clustered flowers. Flowers bright yellow. Pods cylindrical, 6–12.5 cm × 0.5 cm, straw-coloured, blackish or brown, with 6–14 seeds, which are cylindrical to cordate, roundedly 4-sided to subtrapeziform or rectangular with rounded ends, 5–7.5 mm × 4–5.5 mm, smooth, wine red, occasionally buff, creamish, black or mottled. The seeds retain their viability for over 2 years. Germination is hypogeal. The growth period is 60–120 days. More than 60 cultivars have been recorded, differing in time of maturity, colour of seed and plant habit. Cross-pollination between cultivars is frequent. The distinction of var. *angularis* for the cultivated plants and var. *nipponensis* (Ohwi) Ohwi & Ohashi for the primitive (wild?) plants is unreal.

Ecology Adzuki bean requires soil temperatures above 16 °C to germinate and 15–30 °C to grow well. The crop tolerates precipitation of 530–1730 mm/year, an annual mean temperature of 8–30 °C and a pH of 5–7.5. Adzuki bean is a quantitative short-day plant.

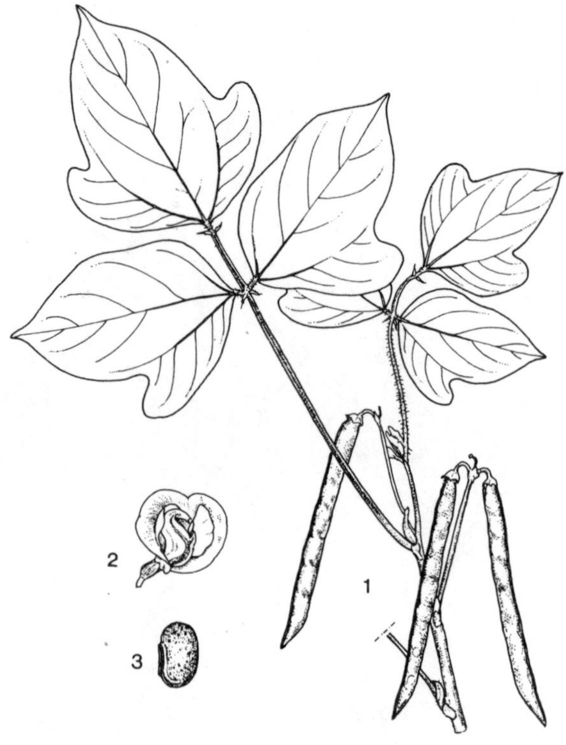

Vigna angularis (Willd.) Ohwi & Ohashi – 1, fruiting branch; 2, flower; 3, seed.

Agronomy Propagation is by seed. Seed is sown 2.5 cm deep in rows 60–90 cm apart, plants 30 cm apart. Seeding rates are 8–30 kg/ha. In Japan, adzuki beans are cultivated in rotation with rice grown in rows. Elsewhere the seed is sown directly in the rice stubble at a high rate, reducing the weed problem.

In Japan, several insect pests, like the adzuki pod worm, the Japanese butter bean borer and cutworms attack the crop. Several fungi and bacteria are known to cause diseases in adzuki bean: bacterial blight, leaf-spot, leaf blotch, rust. Bean weevil, *Callosobruchus chinensis*, can attack the seed during storage.

In general, the pods do not shatter readily; the crop can be harvested with a mower or bean harvester. Some pods are thin and seed may germinate in the pods in wet conditions. For hay, adzuki bean should be cut when the pods are half mature. For seed, cutting is done when all pods are mature. Yield of seeds can be 90–2500 kg/ha. In Taiwan, it is 1450 kg/ha, in Japan 1900 kg/ha, in Kenya 500–600 kg/ha and in New Zealand 1340–2240 kg/ha.

Genetic resources and breeding Unknown in

the wild, germplasm of adzuki beans must be collected from the regions of cultivation in India, China and Japan. In Japan, breeding has resulted in several high-yielding cultivars like 'Benidainagon', 'Chungjupat', 'Hatsune chouzou' and 'Erimoshozu'.

Prospects Adzuki bean is a suitable crop for the subtropics and high altitudes in the tropics. Although it is a valuable pulse, the potential of adzuki bean against erosion should not be overlooked. Its tolerance to drought, frost, heat and virus diseases makes it an interesting crop, deserving more attention.

Literature |1| Duke, J.A., 1981. Handbook of legumes of world economic importance. Plenum Press, New York and London. pp. 288–291. |2| Sacks, F.M., 1977. A literature review of Ph. angularis – the adzuki bean. Economic Botany 31(1):9–15.

C.C.C.M. van Oers

Vigna dalzelliana (O. Kuntze) Verdc.

Kew Bulletin 24: 558 (1970).

LEGUMINOSAE

$2n = 22$

Synonyms *Phaseolus pauciflorus* Dalzell (1851), non G. Don (1832) nec Bentham (1837), *Phaseolus dalzellianus* O. Kuntze (1891), *Phaseolus dalzellii* Cooke (1902).

Vernacular names Indochina: a re (Lâm Dông), re bo nhui (Thuân Hai). India: mugavaine, mugavel, ranmug.

Origin and geographic distribution The species is found in Thailand, Cambodia, Laos, Vietnam, the Philippines and India.

Uses The seeds are used in the same way as green gram (*Vigna radiata* (L.) Wilczek). The whole plant is used as fodder and is excellent to conserve soil.

Botany Slender climbing or creeping annual herb, 4–5 m long, with glabrescent stems rooting at the nodes. Leaves trifoliolate, petiole 1–6 cm long; leaflets broadly ovate, 2–5 cm × 1.5–3 cm, acuminate, entire or faintly lobed. Inflorescences in axillary racemes, 4–10 cm long, with 2–5 flowers; bracteoles longer than the calyx; corolla 1 cm long, yellow. Pods subcylindrical flattened, 3–5 cm × 0.3 cm, with 4–10 seeds, which are reniform, 2–3 mm × 2 mm, brown.

In India, flowering is in September–October; fruiting in October–November. Two varieties are distinguished: var. *dalzelliana* (leaflets ovate) and var. *elongata* Nguyên Van Thuân (leaflets linear to linear-lanceolate, in Indochina).

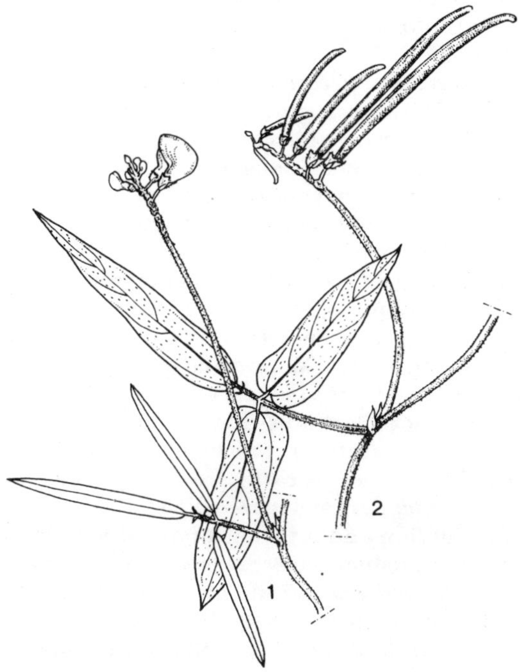

Vigna dalzelliana (O. Kuntze) Verdc. – 1, flowering branch; 2, fruiting branch.

The species much resembles *V. minima* (Roxb.) Ohwi & Ohashi and *V. umbellata* (Thunb.) Ohwi & Ohashi var. *gracilis* (Prain) Mar. Masch. & Stain.

Ecology Var. *dalzelliana* occurs especially on burnt places in forests, on black soils, up to altitude 2000 m in the tropics. Var. *elongata* occurs mainly in gorges on sandy soils.

Agronomy *V. dalzelliana* is a plant growing in the wild as a component of the ground flora of monsoon forests.

Prospects *V. dalzelliana* can be grown to conserve soils.

Literature |1| Babu, C.R., Sharma, S.K. & Johri, B.M., 1985. Leguminosae – Papilionoidae. Tribe – Phaseoleae. Bulletin of the Botanical Survey of India 27(1–4):14. |2| Nguyên Van Thuân, 1979. Flore du Cambodge, du Laos et du Viêt-Nam 17:193–194.

P.C.M. Jansen

Vigna mungo (L.) Hepper

Kew Bulletin 1956: 128 (1956).
LEGUMINOSAE
$2n = 22$

Synonyms *Phaseolus mungo* L. (1767), non Roxb. (1832) & auct. mult.

Vernacular names Black gram, urd bean (En). Haricot mungo, ambérique (Fr).

Origin and geographic distribution A primary gene centre of black gram is found in India and a secondary centre in Central Asia. In South-East Asia, black gram is cultivated in northern Malaysia, the Philippines, Thailand and Burma. It is also cultivated in Bhutan, Nepal, Bangladesh, Afghanistan, Iran, Kenya, Malawi and the United States but it is a major crop in India only.

Uses Black gram is used as a pulse, direct or in various preparations for man to eat. Small amounts are used as cattle feed. It has some use as a green manure and in medicine.

Production and international trade In India, the major producer, average production in the period 1980–1985 was 1.07 million t/year from an area of 2 996 000 ha, respectively. In India, black gram is grown for local use. Thailand is the second producer and dominates world trade.

Properties Dry seeds contain per 100 g edible portion: water 10 g, protein 22–24 g, fat 1–2 g, carbohydrates 56–60 g, fibre 0.9 g, ash 3.2 g. The energy content averages 1445 kJ/100 g. Seed weight varies between 1.5 and 4 g/100 seeds.

Botany An erect hairy annual, 30–100 cm tall, sometimes twining; stem diffusely branched from the base, furrowed. Leaves trifoliolate, petiole 6–20 cm long; leaflets ovate or rhombic-ovate, 4–10 cm × 2–5 cm, entire, acuminate. Flowers small and yellow, in axillary racemes on initially short but later elongating (up to 18 cm) peduncles; bracteoles linear to lanceolate, exceeding the calyx; corolla yellow with spirally coiled keel with a terminal horn-like appendage. Pods subcylindrical, 4–6 cm long, with long hairs and short hooked beak; seeds 4–10 per pod, ellipsoid, with square ends and concave, raised hilum, usually black or mottled. Germination in 7–10 days. Vegetative growth up to 50–60 days from sowing and maturity in 90–120 days. Germination is epigeal.
Three varieties are distinguished:
- var. *mungo*, with large black-seeded and early-maturing types;
- var. *viridis* Bose, with greenish dull or shining glossy seeds and late maturing types;
- var. *silvestris* Lukoki, Maréchal & Otoul, a wild type occurring in open shrubland (scrub) vegetation in the Western Ghats and northern hilly tracts of Maharashtra, India, in semihumid tropical habitats.

Vigna mungo (L.) Hepper – 1, fruiting branch; 2, flower; 3, seed.

The wild plant is viny and hairier with denser umbels and smaller seeds than cultivated types with prominent raised hilum. For the cultivated forms (the first two varieties), a classification into cv. groups would be more appropriate.

Ecology Black gram is basically a warm-season crop but it is grown in both summer and winter in India, depending on the absence of frost. Heavier soils like black-cotton soils are preferred. It is drought-resistant but intolerant of frost and prolonged cloudiness. Black gram tolerates a precipitation of 530–2430 mm/year, an annual mean temperature of 8–28 °C and a pH of 4.5–7.5. Optimum temperature is 25–35 °C and yields are reasonable with a precipitation of 650–900 mm/year.

Agronomy Black gram is propagated by broadcasting or by sowing in rows. It does not require thorough field preparation; rough tillage suffices with weeding once or twice. The crop is mainly rain-fed and fertilizer application is not common. Serious diseases are leaf-spot (caused by *Cerco-*

spora sp.) and yellow mosaic virus. Pests include aphids, bugs and weevils. In storage, the seeds are attacked by bruchids (*Callosobruchus chinensis*). Black gram should be harvested before the pods are fully ripe, to prevent shattering. Seeds are processed in the form of split seeds (dhal). Yield of dry seed averages 340–560 kg/ha but it can reach 1500 kg/ha.

Genetic resources and breeding Genetic diversity is narrower than in green gram. The AVRDC in Taiwan maintains a collection of 200 entries (1980). About 1900 accessions are maintained by the National Bureau of Plant Genetic Resources, Delhi, at its various research stations. Breeding programmes to improve this nutritious pulse have not been commensurate with its role in the Indian diet. Systematic and planned work for improvement started in India in the early 1940s. Over 40 cultivars, both erect and spreading types, are grown in different agroclimatic regions of India. Sources of resistance to most current diseases are available and progress is being made by AVRDC to incorporate those traits.

Prospects Black gram is a promising crop for South-East Asia. It is nutritious and its ecological suitability is wide. Diversity of germplasm needs to be exploited in the South-East Asian countries.

Literature |1| Anonymous, 1976. The wealth of India. CSIR, New Delhi. Vol. 10. pp. 476–484. |2| Duke, J.A., 1981. Handbook of legumes of world economic importance. Plenum Press, New York and London. pp. 291–293.

R.K. Arora & Shri S. Mauria

Vigna radiata (L.) Wilczek

Flore du Congo Belge et du Ruanda-Urundi 6: 386 (1954).

LEGUMINOSAE

$2n = 22$

Synonyms *Phaseolus radiatus* L. (1753), *Phaseolus aureus* Roxb. (1832).

Vernacular names Mung bean, green gram, golden gram (En). Haricot mungo, haricot doré (Fr). Indonesia: kacang hijau, arta ijo. Malaysia: kacang hijau. The Philippines: mongo, balatong. Burma: pè-di-sien. Cambodia: sândaèk ba:y. Laos: thwàx khiêw, thwàx ngo:k, thwàx sadê:k. Thailand: thua khieo, thua thong. Vietnam: dâu xanh, dâu chè.

Origin and geographic distribution Mung bean originated in India or the Indo-Burmese region where it has been cultivated for several millennia. It spread in early times to most other Asian countries and more recently to other continents. Despite its present wide distribution, mung bean never became a major commercial crop outside Asia. In most South-East Asian countries, mung bean ranks among the three main grain legumes.

Uses The dried beans are prepared by cooking or milling. They are eaten whole or split (dhal). The seeds or the flour may enter a variety of dishes like soups, porridge, snacks, bread, noodles and even ice-cream. Mung bean starch is extensively used for starch noodles; mung bean protein to fortify cereal flours. Both fractions can be separated by air blowing. Most popular as a fresh vegetable in oriental cooking are sprouted mung beans. Crop residues are a useful fodder. Mung bean is sometimes specifically grown for hay, green manure or cover crop.

Production and international trade World production of mung beans in 1980s is estimated to be 2 million t from 4 million ha. India is by far the main producer (1.2 million t from 2.8 million ha), followed by Thailand (0.25 million t from 0.4 million ha) and Indonesia (0.20 million t from 0.3 million ha). With mung bean losing its image as 'poor man's meat' through high market prices, there is ample scope to increase production to meet the growing domestic demand in most South-East Asian countries. In Indonesia, harvested area doubled and domestic production tripled in 1975–1985. International trade is dominated by Thailand, which exports half of its production of 250 000 t/year, mainly to Japan, Taiwan, the Philippines, Malaysia and Singapore, but also to Europe and United States.

Properties Mung bean has a higher digestibility and causes less flatulence than most other pulses, making it suitable for children and older people. Per 100 g edible portion, dry mung beans have an energy content of 1430 kJ and contain: water 10 g, protein 22 g, fat 1 g, carbohydrates 60 g, fibre 4 g and ash 3 g. Variation in protein content (17–26 %) is more influenced by environment than by genotype. Mung protein is deficient in methionine and cystine, but rich in lysine, making it an excellent complement to rice. Mung beans are a good source of minerals, provitamin A and vitamin B complex, while bean sprouts are rich in ascorbic acid (vitamin C) too. Mung bean is low in antinutritional factors. Heating and sprouting denatures growth inhibitors. Seed weight varies between 1.5 and 8.5 g/100 seeds.

Description An erect or semierect, sometimes twining, herbaceous annual, 25–130 cm tall.

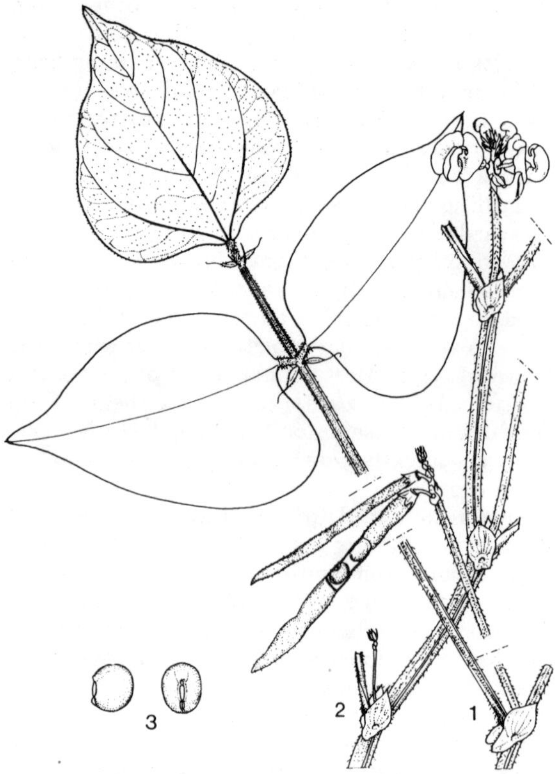

Vigna radiata (L.) Wilczek – 1, fruiting branch; 2, flowering branch; 3, seeds.

Branching starts from lower and intermediate nodes (rarely from cotyledons or unifoliolate leaves). First two leaves opposite and simple, subsequent leaves alternate and trifoliolate; leaflets ovate to deltoid, 5–18 cm × 4–15 cm, usually entire. Flowers are large, diameter 1–2 cm, greenish to bright yellow, self-fertile, borne in clusters of 5–25 on axillary racemes 2–20 cm long. Pods spreading and pendent, cylindrical, up to 15 cm long, usually straight, pubescent or glabrous, black or tawny brown with up to 20 globose to ellipsoid seeds. Seeds green or yellow, occasionally brown or blackish, with dull or glossy lustre (associated with remains of pod wall); hilum flat and white. Germination is epigeal.

Growth and development Mung bean is a short-duration crop, usually flowering within 30–70 days and maturing within 60–120 days of sowing. Flower abscission is prevalent and may reach 90 %. Mung bean is usually determinate but because the inflorescences remain meristematic and may redevelop flowers after a period of adverse conditions, it flowers and fruits over a period of several weeks. Green leaves, open flowers, green pods and ripe pods occur simultaneously on the same plant. A large part of the dry matter accumulated during seed filling may still be partitioned to vegetative parts. So rapid senescence (shedding of the leaves uniformly from the plant) does not occur. Although self-pollination is the rule and cleistogamy common, outcrossing up to 5 % may occur. Flowers are usually pollinated during the night, before they open early in the morning. It takes 3–4 weeks from flower opening to mature pod.

Other botanical information Mung bean is morphologically similar to black gram (*Vigna mungo* (L.) Hepper) and they are sometimes considered variants of the same species. Chemotaxonomy and cytogenetics support distinction but the debate still continues. Both species are sometimes considered to be domesticates from the same wild forms (*Vigna sublobata*, based on *Phaseolus sublobatus* Roxb.). The cultivated forms of mung bean are now usually grouped as *V. radiata* var. *radiata*, though a classification into cultivar groups would be more appropriate. The wild forms are usually classified into two botanical varieties:
- var. *sublobata* (Roxb.) Verdc. is smaller in all parts than var. *radiata*, and occurs in India, Sri Lanka, South-East Asia, Queensland, Madagascar and East Africa;
- var. *setulosa* (Dalzell) Ohwi & Ohashi has large, almost orbicular stipules and dense long hairs on the stem. It occurs in India, China, Japan and Indonesia.

Ecology Mung bean is a short-day plant. Cultivars differ markedly in sensitivity but most genotypes show quantitative short-day responses, flower initiation being delayed by increases in the photoperiod. Qualitative responses (no flower initiation if photoperiod longer than a certain critical value) occur, while absolute day-neutrality has yet to be confirmed. Mung bean is a warm-season crop and grows within a mean temperature range of 20–40 °C, the optimum being 28–30 °C. It can therefore be grown in summer and autumn in warm temperate and subtropical regions and at altitudes below 2000 m in the tropics. The crop is susceptible to waterlogging but withstands drought well, by curtailing the period from flowering to maturity. Water requirement is 200–300 mm for the growing season. The crop does best on well drained loam or sandy loam, with pH 5.5–7.0. There is little information on tolerance to aluminium toxicity. Mung bean nodulates readily with *Rhizobium* strains from the cowpea cross-inoculation group. Because those strains are rather common, there is little response to inoculation.

Propagation Propagation is by seed. There is no seed dormancy. Seeds may sprout in the pod under very humid conditions. Seed is broadcast or dibbled in hills or in rows. Mung bean is grown mainly on smallholdings, often as mixed crops or intercrops. Associated crops are usually of longer duration than the mung bean (sugar-cane, cotton, sorghum) but sometimes of shorter duration (Madura maize, Indonesia). To make use of a short cropping period, short-duration mung bean is often relay-cropped.

Husbandry In the major area of cultivation, the monsoon tropics, mung bean is mainly grown as rainy season crop on dryland or as dry-season crop after the monsoon in rice-based systems on wetland, making use of residual moisture or supplementary irrigation. In some areas where adequate early rains occur, an early-season crop before the monsoon can be grown. With the newer cultivars ripening in 60–75 days, maximum yields are obtained at plant densities of 300 000–400 000 plants/ha. The later-maturing traditional cultivars generally need wider spacing. Usually no fertilizers are applied to mung bean. It uses residues from applications to the main crops in the system, though it responds well to P_2O_5 at a rate of 20–80 kg/ha. Nutrient removal is N 40, P 4, K 12, Ca 1, S 2 and Mg 2 kg per tonne of dry seed. The loss is usually much higher with crop residues for fodder. Over the centuries, mung bean's adaptation to stable performance in marginal environments has resulted in a low yield potential, which limits responsiveness to better environments and improved cultural practices.

Diseases and pests The most serious fungal diseases are leaf-spot (*Cercospora canescens*) and powdery mildew (*Erysiphe polygoni*). Less serious are scab (*Elsinoe iwatae*), anthracnose (*Colletotrichum lindemuthianum*) and rust (*Uromyces* spp.). Serious bacterial diseases are blights caused by *Xanthomonas* and *Pseudomonas*. Mung bean suffers from several virus diseases but, except for mung bean yellow mosaic virus (MYMV), they are not well described. Beanflies (*Ophiomyia phaseoli* and related species) can cause serious stand reductions at the seedling stage and much economic damage is done by pod-borers (*Heliothis* spp., *Etiella zinckenella*, *Maruca testulalis*) and pod-suckers such as the green stink bug (*Nezara viridula*). A serious storage insect is the seed weevil (*Callosobruchus* spp.), which destroys the whole seed-lot, if uninterrupted.

Harvesting Generally by 2–5 hand-pickings at weekly intervals. Harvesting is the most expensive single operation in growing mung beans. Short-duration cultivars, which ripen more uniformly, may be processed as whole plants on small rice threshers. Cultivars differ markedly in harvesting efficiency, depending on position (above or within canopy) and size of pods.

Yield Yields under monocropping are low, 300–700 kg/ha in the major producing countries. In trials, yields higher than 3 t/ha have been reached.

Handling after harvest Hand-picked pods are dried in the sun. Spontaneous shattering can be speeded up by beating with a stick or by trampling. Seed is cleaned by screening and winnowing. Properly dried mung beans maintain high viability over a long period. Seed stored by small farmers for sowing is often of poor quality through bruchid damage. For long-term storage and overseas shipments, seeds are usually fumigated.

Genetic resources Holders of the largest germplasm collections are AVRDC, Shanhua, in Taiwan (6000 accessions), United States Department of Agriculture, Georgia and Colorado, in the United States (3500), Punjab Agricultural University, Ludhiana, in India (3000), and CAAS, Beijing, China (3000). There are several smaller but useful collections at research institutes in South-East Asia: FCRI (Bangkok) in Thailand, BORIF (Bogor) and MARIF (Malang) in Indonesia and IPB (Los Baños) in the Philippines. Internationally, the emphasis is now on evaluation and documentation work to improve user's access to the collections.

Breeding In many areas that traditionally grow mung beans, farmers still grow old landraces, mixtures of homozygous genotypes that are well equipped to adjust to changing environmental conditions and give stable but seldom excellent yields. Even recently, many cultivars have been developed from those landraces by simple pureline selection and identification of the outstanding genotypes in those mixtures. Those well adapted local selections form an excellent starting point for hybridization programmes. The challenge in breeding is to overcome the genetic limitations deriving from centuries-long adaptation of mung bean to marginal environments. The traditional late robust types have been replaced by new types useful for short seasons and multiple cropping systems with mung bean occupying the land for short periods between major crops. These are short compact plants with a high harvest index, with reduced photoperiodic sensitivity and with a more uniform maturity. Many modern cultivars have been released in the major producing countries that also have improved resistance to major pests and diseases. Sources of resistance have been identified in germplasm of mung bean and related species. Black gram (*Vigna mungo*) shows, among the Asiatic *Vigna* species, most promise for interspecific hybridization with mung bean. Little attention is being paid to protein content and quality, because much more impact on total protein production is to be expected from increasing yields.

Prospects Demand for food legumes will continue to grow. Short-duration mung bean is suitable to increase the cropping intensity of agricultural land. New developments in food technology allow expansion of use of mung bean. Bean sprouts are gaining popularity as a winter vegetable in many temperate areas. Research on production should focus on reliable yield and good seed quality (weathering resistance). Processing research should focus on diversification and quality of mung bean products.

Literature |**1**| AVRDC, 1979–1986. Progress report summaries. AVRDC, Taiwan. |**2**| AVRDC, 1988. Proceedings of the Second International Mungbean Symposium. AVRDC, Taiwan (in press). |**3**| Cowell, R. (Editor), 1978. Proceedings of the First International Mungbean Symposium. AVRDC, Taiwan. 262 pp. |**4**| Lawn, R.J. & Ahn, C.S., 1985. Mung bean (Vigna radiata (L.) Wilczek/ Vigna mungo (L.) Hepper). In: Summerfield, R.J. & Roberts, E.H. (Editors): Grain legume crops. Collins, London. pp. 584–623. |**5**| Morton, J.F., Smith, R.E. & Poehlman, J.M., 1982. The mungbean. College of Agricultural Sciences, Department of Agronomy and Soils, University of Puerto Rico, Mayaguez. Special Publication. 136 pp. |**6**| Kay, D.E., 1979. Food legumes. Tropical Products Institute, London. Crop and Product Digest No 3. pp. 273–292.

J.S.Siemonsma & Arwooth Na Lampang

Vigna subterranea (L.) Verdc.

Kew Bulletin 35(3): 474 (1980).

LEGUMINOSAE

$2n = 22$

Synonyms *Voandzeia subterranea* (L.) Thouars (1806).

Vernacular names Bambara groundnut (En). Voandzou (Fr). Indonesia: kacang Bogor.

Origin and geographic distribution Origin is most probably northern Nigeria and Cameroon. The species is mainly cultivated throughout the drier areas of tropical Africa but has also spread to America, Australia, Central Asia, Indonesia, Malaysia and the Philippines.

Uses Roasted or cooked (ripe pods in salt water) seeds are eaten as a delicacy. Seeds and young pods are added to spicy soups. The main use in Africa is as a pulse, for instance to prepare a porridge.

Production and international trade World production is estimated at 330 000 t/year, half of which is produced in West Africa. There are no reliable statistics for South-East Asia, where production and consumption are local.

Properties Per 100 g edible portion seeds contain: water 11 g, protein 18 g, fat 6 g, carbohydrates 62 g, fibre 5 g, ash 3 g. The energy content averages 1540 kJ/100 g. As in other leguminous seeds, the sulphur-containing amino acids, cystine and methionine, are limiting, whereas the content of lysine is high. Seed weight varies between 50 and 75 g/100 seeds.

Botany Bunch-shaped to spreading annual herb with creeping much branched stems. Roots with lobed nodules. Leaves trifoliolate, glabrous, with erect grooved petioles up to 30 cm long; leaflets elliptic to oblanceolate, up to 8 cm × 4 cm. Racemes with (1–)2(–3) whitish-yellow flowers. Fruits subterranean, subspherical, diameter about 2.5 cm, usually with 1 white, yellow, red-blackish or variously mottled seed. Germination is hypogeal, taking 7–15 days. Flowering starts 30–55 days after sowing. Self-pollination is usual. The pod develops in the first month after fertilization. Then the seed expands during the following 10 days. Seeds are mature when the parenchymatous layer surrounding the embryo has disappeared and brown patches appear on the outside of the pod. Maturity is reached 90–150 days after sowing.

Ecology Bambara groundnut, a short-day plant, is cultivated in the tropics at altitudes up to 1600 m. A frost-free period of at least 3.5 months is necessary. The plant prefers bright sunshine and average day temperatures between 20 and 28 °C. The crop can be cultivated successfully in areas with a rainfall of 600–750 mm/year but optimum yields require 900–1200 mm/year. The plant will grow on any well drained soil but light sandy loams with a pH of 5.0–6.5 are most suitable.

Agronomy Bambara groundnut performs best on a deeply ploughed field with a fine seed-bed, which allows the plant to bury its peduncles with fertilized flowers. Average seed rates are 25–75 kg/ha. Recommended spacing in pure stand in the flat is 10–15 cm apart in single rows at a distance of

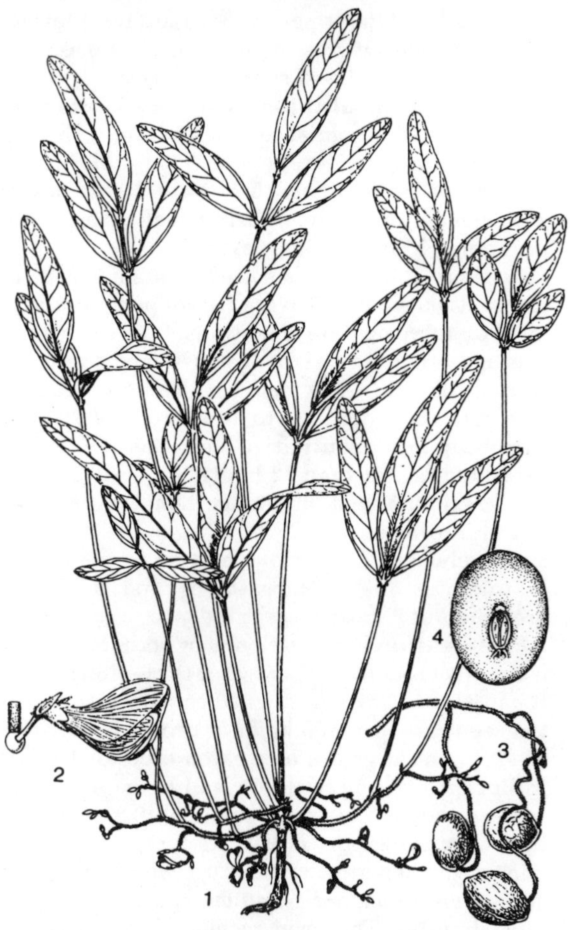

Vigna subterranea (L.) Verdc. – 1, habit of flowering plant; 2, flower; 3, fruits; 4, seed.

45 cm or 20 cm apart in double rows on flat-topped ridges 90 cm apart. As an intercrop, bambara groundnut is combined with cereals, root and tuber crops, and other legumes. After a well fertilized preceding crop, a maintenance dressing with superphosphate is recommended and soils of low fertility require an application of compound fertilizer. The crop should be weeded 2 weeks after emergence and again before the clusters join. In general, pests and diseases do not reduce yields seriously. Harvesting is by hand-pulling when the leaves turn yellow and wither. Since pods may break from the plant on lifting, gleaning is useful. Seed yield in Africa averages 650–850 kg/ha with large differences between countries. For South-East Asia, no statistics on yields are available. Seeds stored in the shell suffer less from deterioration and infestation by insects than shelled seeds.

Genetic resources and breeding The largest collection of germplasm is held by the IITA in Nigeria. So far breeding is for high yields.

Prospects Bambara groundnut is a promising crop for semiarid areas, since it tolerates drought and poor soils better than many other crops.

Literature |**1**| Adenaike, B.O. (Editor), 1978. Bambara groundnut (Voandzeia subterranea Thouars): abstracts of world literature 1900–78. IITA, Ibadan, Nigeria. 55 pp. |**2**| Linnemann, A.R., 1987. Bambara groundnut (Vigna subterranea (L.) Verdc.) – a review. Abstracts on Tropical Agriculture 12(7):9–25.

A.R. Linnemann

Vigna umbellata (Thunb.) Ohwi & Ohashi

Japanese Journal of Botany 44: 31 (1969).

LEGUMINOSAE

$2n = 22$

Synonyms *Phaseolus calcaratus* Roxb. (1832), *Vigna calcarata* (Roxb.) Kurz (1876), *Azukia umbellata* (Thunb.) Ohwi (1953).

Vernacular names Rice bean (En). Haricot riz (Fr). Indonesia: kacang uci. Malaysia: kacang sepalit. The Philippines: anipai, kapilan, pagsei. Burma: pé-yin. Cambodia: sândaêk ângkât miëhs, sândaêk riech mieh. Laos: thwâx la:ng tê:k, thwàx sadê:t pa:x, thwàx phi. Thailand: thua daeng, thua pae, ma pae. Vietnam: dâu gao.

Origin and geographic distribution Rice bean is a native of South and South-East Asia. It is most widely cultivated in China, Korea, Japan, India, Burma, Indonesia, Malaysia, Fiji, the Philippines and Mauritius, and to a limited extent in the tropical parts of all continents.

Uses Rice beans are usually boiled and eaten with rice or instead of rice. The young pods, leaves and sprouts are used as vegetable. The whole plant is used as fodder, as a cover crop, as green manure and as living hedge. In Perak (Malaysia), the leaves are used with rice flour in a poultice applied to the abdomen for stomach ache.

Production and international trade Rice bean rarely enters international trade but is extensively grown for food in Asia and the Pacific islands. In 1975, Japan imported 12 000 t, 7000 t from Thailand, 3000 t from China and 2000 t from Burma.

Properties Per 100 g edible portion dry seeds contain: water 13.3 g, protein 20.9 g, fat 0.9 g, carbohydrates 64.9 g, fibre 4.8 g, ash 4.2 g. The energy content averages 1373 kJ/100 g. Seed weight varies between 8 and 12 g/100 seeds.

Description Annual vining herb with erect, suberect or flexuose stem, 30–75 cm tall, usually clothed with fine, deciduous, deflexed hairs; vines grooved, 1–3 m long. Leaves trifoliolate, stipules lanceolate, petioles 5–10 cm long; leaflets broadly ovate to ovate-lanceolate, 5–10 cm × 2.5–6 cm, membranous, subglabrous, usually entire, sometimes trilobed. Inflorescences erect axillary racemes, 5–10 cm long, with 5–20 flowers, peduncles up to 20 cm long; flowers bright yellow, 2–3 together, diameter up to 2 cm, with large bracteoles. Pods long and slender, partly falcate, 6–13 cm × 0.5 cm, glabrous, with 10–16 seeds, which are oblong to strongly elongate, subtrapezoidal, 5–10 mm × 2–5 mm, smooth, dark red, green, yellow, brown, black, speckled or mottled.

Growth and development Germination is hypogeal. Seedlings grow vigorously and establish themselves early. The plants can smother weeds. The crop matures in 60–150 days. In the Philippines, time to flowering averages 64 days, to maturity 92 days; in India, early-maturing types behave likewise; late types ripen in 130–150 days. Flowers are self-fertile but cross-pollination occurs as well. All pods mature almost simultaneously.

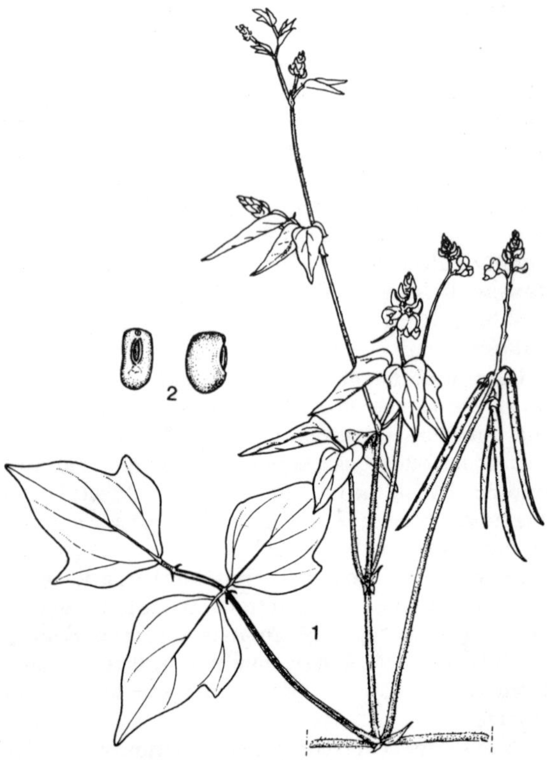

Vigna umbellata (Thunb.) Ohwi & Ohashi – 1, flowering and fruiting branch; 2, seeds.

Other botanical information Two botanical varieties are distinguished: var. *umbellata*, the cultivated forms, and var. *gracilis* (Prain) Maréchal, Mascherpa & Stainier, the original wild forms with slender branchlets, narrow leaflets and long peduncles. Many cultivars exist.

Ecology Rice bean is suited to humid tropical lowlands but some cultivars are adapted to subtropical or subtemperate conditions. It grows best at average temperatures of 18–30 °C and prefers a rainfall of 1000–1500 mm/year. The crop can be grown in the tropics up to an altitude of 2000 m. Rice bean is a quantitative short-day plant. The threshold is less than 12 hours. In West Bengal, the maximum and minimum temperatures for flower initiation are 25–26 °C and 10–12 °C, respectively. Rice bean grows best on fertile loams. It tolerates high temperatures and moderate drought, but is frost-susceptible.

Propagation Propagation is by seed. Rice bean is usually broadcast, after two or three ploughings. It can also be sown in rows 90 cm apart. In India, the seed rate is 40–50 kg/ha if grown for seed or 60–70 kg/ha if grown as a catch crop for fodder. In Burma, average seed rate is 21 kg/ha. In north-eastern India, it is grown under shifting cultivation with maize and millet. Its forage quality is improved when grown in mixture with the annual grass *Pennisetum pedicellatum* Trin.

Husbandry The crop receives little care. In India, superphosphate at 50–60 kg/ha is recommended. In Burma, rice bean is usually grown in rotation with rice.

Diseases and pests Rice bean is fairly pest-free. Powdery mildew, rust and cucumber mosaic virus can attack the crop in the Philippines. Nematode problems are reduced by flooding and so rotation with rice is recommended.

Harvesting The viny habit and the shattering of pods make rice bean difficult to harvest. Harvesting in the morning when the pods are moist reduces the losses. If grown for fodder, rice bean should be harvested when the pods are half-developed, since the leaves drop easily when the plant reaches maturity.

Yield Average seed yield is 200–300 kg/ha, though in West Bengal, with good crop management, yields up to 2240 kg/ha have been obtained. Yield averages 420–840 kg/ha in Burma and 500–800 kg/ha in Papua New Guinea. Forage yield is 2200–3500 kg/ha.

Handling after harvest The seeds are dried in the sun and threshed by hand. Usually the seeds are not affected by storage insects.

Genetic resources and breeding Germplasm collections are available in India and Taiwan (AVRDC), but more collections are needed. The breeding of quick-maturing day-neutral high-yielding and non-shattering erect cultivars that are nematode-resistant are the most urgent research needs. Moreover, all agronomical aspects need investigation, e.g. time of sowing, plant density and fertilizer requirements.

Prospects Development is handicapped by low average yields and by pods shattering easily, making economic harvesting difficult. Its tolerance of high temperatures and humidities, its adaptation to heavy soils, its quick growth, its resistance to pests and diseases, and its nutritious seeds make rice bean a valuable crop that deserves testing throughout the tropics. There has been increasing interest in India in the development of rice bean as a fodder crop and now as a grain crop, and also in West Africa, where it is less susceptible to pests and diseases than many other grain legumes.

Literature |1| Arora, R.K., Chandel, K.P.S., Joshi, B.S. & Pant, K.C., 1980. Rice bean: tribal pulse of eastern India. Economic Botany 34(3):260–263. |2| Chandel, K.P.S., Arora, R.K. & Pant, K.C., 1988. Rice bean – a potential grain legume. NBPGR, New Delhi, India. Scientific Monograph No 12. 60 pp. |3| Kay, D.E., 1979. Food legumes. Tropical Products Institute, London. Crop and Product Digest No 3. p. 348–354. |4| Mukherjee, A.K. & Roquib, M.A., 1980. Herbage and grain yield potentiality of rice bean (Vigna umbellata (Thunb.) Ohwi & Ohashi) – a review. Forage Research 6:165–170.

C.C.C.M. van Oers

Vigna unguiculata (L.) Walp.

Repertorium botanices systematicae 1: 779 (1843).

LEGUMINOSAE

$2n = 22$

Synonyms

- cv. group Unguiculata: *Dolichos unguiculatus* L. (1753), *D. sinensis* L. (1754), *Vigna sinensis* (L.) Hassk. (1844), *Vigna unguiculata* (L.) Walp. ssp. *unguiculata* (1970);
- cv. group Biflora: *Dolichos biflorus* L. (1753), *D. catjang* Burm. f. (1768), *Phaseolus cylindricus* L. (1754), *Vigna catjang* (Burm. f.) Walp. (1839), *Vigna unguiculata* (L.) Walp. ssp. *cylindrica* (L.) van Eseltine (1931);
- cv. group Sesquipedalis: *Dolichos sesquipedalis* L. (1763), *Vigna sesquipedalis* (L.) Fruhw. (1898), *Vigna sinensis* (L.) Hassk. ssp. *sesquipedalis* (L.) van Eseltine (1931), *Vigna unguiculata* (L.) Walp. ssp. *sesquipedalis* (L.) Verdc. (1970).

Vernacular names

- cv. group Unguiculata: (common) cowpea, black-eye bean, southern pea (En). Haricot dolique, niébé (Fr). Indonesia: kacang tunggak. Malaysia: kacang bol, kacang toonggak, kacang merah. The Philippines: batong, kibal, otong. Cambodia: sândaêk kâng, sândaèkângkuy. Laos: thwàx do. Thailand: tua dam. Vietnam: dôu den, dôu trang, dôu tua;
- cv. group Biflora. (Catjang) cowpea, sowpea (En). Dolique (Fr). Indonesia: kacang merah, kacang peudjit, kacang tunggak. Cambodia: sândaêk sâ, sândaêk khmau, sândaêk krâhâm. Laos: thwàx siênx. Thailand: thua khaao, thua rai, po-thoh-saa. Vietnam: dâu ca, dâu trang, dâu do;
- cv. group Sesquipedalis. Yard-long bean, asparagus bean (En). Dolique asperge, haricot kilomètre (Fr). Indonesia: kacang panjang, kacang tolo, kacang belut. Malaysia: kacang panjang, kacang bêlut. The Philippines: sitao, hamtak, banor. Cambodia: sândaèk troeung. Thailand: tua fak yaow, tua phnom. Vietnam: dâu dûa, dâu giai áo.

Origin and geographic distribution *V. unguiculata* originated in Africa, though where the crop was first domesticated is uncertain. Two centres of diversity appear to exist for the species, which contains wild and cultivated forms: one in West Africa (for cv. group Unguiculata) and another in India and South-East Asia (for cv. group Biflora and cv. group Sesquipedalis). Common cowpea is widely distributed throughout the tropics and subtropics (30 °N–S), especially in Africa. Outside Africa, it is also grown in Asia, especially India, Australia, the Caribbean, the southern United States and the lowland and coastal areas of South and Central America. Catjang cowpea is cultivated mainly in India and Sri Lanka, and, to some degree, in South-East Asia. Yard-long bean is mostly cultivated in India, Bangladesh and South-East Asia, and Oceania, but it has spread widely throughout the tropics as a minor vegetable crop.

Uses

- Cv. group Unguiculata. Cultivated for the seeds (shelled green or dried), the pods or leaves that are consumed as green vegetables or for pasture, hay, silage and green manure. In Africa, where they are the preferred food legume, they are consumed in 3 basic forms: (1) cooked together with vegetables, spices and often palm oil, to produce a thick bean soup, which accompanies the staple food (cassava, yams, plantain); (2) decorticated and ground into a flour and mixed with chopped onions and spices and made into cakes that are either deep-fried (akara balls) or (3) steamed (moin-moin). In India, the common cowpea is used mostly as a pulse, either whole or as dhal. Leaves may be boiled, drained, sun-dried and then stored for later use.

- Cv. group Biflora. In India and Sri Lanka, it is grown for seed and as a vegetable. Tender green pods are consumed as green vegetable. Dried seeds are used whole or split. It makes excellent forage. It is used to make hay and is often mixed with maize or sorghum for silage and for green manure.
- Cv. group Sesquipedalis. Yard-long bean is grown for its succulent young pods and sometimes for its leaves as vegetable. Dry seeds are cooked with meat and fish. Green plants are used as fodder or as a green manure.

Production and international trade Worldwide production in 1981 was estimated conservatively at 2.27 million t from 7.7 million ha. Cowpeas are grown extensively in 16 African countries, Africa producing two thirds of the total. Nigeria and Niger produce half the world crop. Brazil produces 26 % of the world total. The estimated area in Asia under different forms of the species is 1 million ha concentrated in India (more than 0.5 million ha), Sri Lanka, Burma, Bangladesh, the Philippines, Indonesia, Thailand, Pakistan, Nepal, China and Malaysia. Most production in South-East Asia is as green vegetable and a limited amount as dry seed. Most production is by smallholders.

Properties Mature seeds contain per 100 g edible portion: water 10 g, protein 22 g, fat 1.4 g, carbohydrates 59.1 g, fibre 3.7 g, ash 3.7 g, calcium 104 mg and small amounts of other nutrients. The energy content averages 1420 kJ/100 g. Lysine content is high, making cowpea an excellent improver of protein quality of cereal grains. Seed weight varies between 10 and 25 g/100 seeds.

Raw young green pods per 100 g edible portion: water 88.3 g, protein 3.0 g, fat 0.2 g, carbohydrates 7.9 g, fibre 1.6 g, ash 0.6 g. The energy content averages 155 kJ/100 g.

Description A prostrate, climbing, erect to suberect, nearly glabrous annual, 0.3–4 m long, with well developed root system. Stems more or less square, slightly ribbed, with nodes usually violet. Stipules prominent, ovate, appendaged. Leaves alternate, trifoliolate, with petiole 5–25 cm long; first two leaflets opposite, asymmetrical, top leaflet symmetrical, ovate, sometimes shallowly lobed, (6.5–)7.0–13.5(–19.5) cm × (3.5–)4.0–9.5 (–17.0) cm. Inflorescences are axillary racemes with several flowers clustered near the top; peduncle (4–)10–17(–32) cm long; rachis contracted, tuberculate; fertile flowers attached to a tubercle carrying abortive flowers, leaving gland-like tissue after being shed; bracts 1 per flower, early deciduous; pedicel short; bracteoles 2, deciduous, obovate, 3–5 mm long; calyx campanulate, lobes 5–7 mm long; corolla with erect or spreading standard, 2–3 cm long, hood-shaped when older, wings 22 mm × 12 mm, keel boat-shaped, 21 × 12 mm; stamens diadelphous (9 + 1); ovary with 12–21 ovules. Pod pendent or erect to spreading, linear, 10–100 cm long. Seeds variable in size and shape, square to oblong, 5–10 mm × 4–8 mm, variously coloured.

Vigna unguiculata (L.) Walp. – 1, fruiting branch; 2, inflorescence; 3, seed.

Growth and development Cowpea seeds can

germinate in 3–4 days under favourable soil moisture and temperature (28 °C). Germination is epigeal. Within 5 days, the cotyledons have lost much of their weight and have begun to abscise. Size of the cotyledons has a direct positive influence on the size of the emerging seedling. The maximum leaf area index (3–4) is achieved between flowering and early pod set in most cultivars, when the crop intercepts the maximum solar radiation. Cowpea can show extreme variation in start and finish of the reproductive period. Some cultivars may start flowering 30 days after sowing and be ready for harvest of dry seed 25 days later; others may take more than 90 days to flower, and take 210–240 days to mature. Most flowers are self-pollinated, though a small proportion of outcrossing occurs, especially in humid climates. Most genotypes of cowpea respond to photoperiod as typical quantitative short-day plants but some genotypes are insensitive to a wide range of photoperiods and warmer circumstances can hasten the appearance of flowers in genotypes that are either sensitive or insensitive to photoperiod. The period of anthesis is characterized by profligate loss of flower buds and open flowers, and afterwards of immature fruits. Pod development is rapid and lasts about 19 days.

Other botanical information Within *V. unguiculata*, the cultivated forms are classified as follows.

- Cv. group Unguiculata (sometimes *V. unguiculata* ssp. *unguiculata*), the common cowpea: spreading, suberect or erect annual, 15–80 cm high; pods 10–30 cm long, pendent (even when young), hard and firm, not inflated when young; seeds usually 6–10 mm long.
- Cv. group Biflora (sometimes *V. unguiculata* ssp. *cylindrica* (L.) van Eseltine), the catjang cowpea: spreading, suberect or erect annual, 15–80 cm high; pods 7.5–12 cm long, erect or ascending, hard and firm, not inflated when young; seeds usually 3–6 mm long.
- Cv. group Sesquipedalis (sometimes *V. unguiculata* ssp. *sesquipedalis* (L.) Verdc.), the yard-long bean: climbing annual, 2–4 m high; pods 30–100 cm long, pendent, more or less inflated and flabby when young; seeds usually 8–12 mm long.

Ecology Cowpea is a quantitative short-day plant. Local populations of cowpeas grown by farmers in West Africa are well adapted so that they start to flower at the end of the rains at a particular locality. The optimum temperature to grow and develop is 20–35 °C. Cowpea can grow in a wide range of soils and does well on acid soils. Excessive soil moisture reduces growth but drought can be tolerated.

Propagation Propagation is by seed. Common cowpea is traditionally grown in Africa intercropped with cereals like pearl millet, sorghum or maize, at wide spacings (total plant population 10 000–20 000 plants/ha). The bulk of production comes from smallholdings. Tillage normally follows the crop with which cowpea is interplanted. Rate of sowing varies with planting method: when sown in rows, 10–40 kg/ha; for broadcasting, 90 kg/ha. Cowpea is grown in many parts of Asia in sole cropping or intercropped with cereals, cotton or sugar-cane, and relay-cropped in standing rice. Cultivation practices differ widely in the region. In southern India, Sri Lanka, the Philippines, Indonesia and Thailand, most of the crop is grown after rice. During this period, rainfall diminishes and later in the season soil moisture is limited. In rain-fed areas of South-East Asia, early-maturing cultivars could be sown in late April–early May, with harvest at the end of June–early July, before rice is transplanted. For catjang cowpea, seed rate is 15–20 kg/ha in sole cropping but lower in intercropping. Yard-long bean is grown on rice bunds or as a backyard crop. It is grown as sole crop or intercropped with maize, sugar-cane or cassava. Near cities, most farmers grow it as a sole crop on a small scale. Seed rates are 25–50 kg/ha. Yard-long bean is established in rows 75–100 cm apart with hill-to-hill spacing of 20–25 cm. Plant population density is 60 000–70 000 plants/ha.

Husbandry Most cowpea crops are rain-fed, a few are irrigated and others use residual moisture in the soil after harvest of a rice crop. Cowpea is particularly well suited for rice-based cropping systems. Two to three weedings during the first 1.5 months after planting are recommended. Losses due to weeds can be 30–65 %. Parasitic weeds (*Striga* spp.), generally associated with continuous cropping of cowpea in Africa, may cause damage too. Plants of yard-long bean are staked when 25–30 days old. In most soils, native *Rhizobium* strains nodulate the plants. Effective cowpea–*Rhizobium* symbiosis fix more than 150 kg/ha of N and supply 80–90 % of the nitrogen the host plant requires. Inoculation may be advantageous, if the crop has not been grown for many years. In general, no fertilizers are applied. However P_2O_5 at 30 kg/ha often improves performance. A starter of N at 30 kg/ha may also be beneficial. Cowpea is commonly incorporated in crop rotations in semiarid, humid and subhumid environments.

Diseases and pests Cowpea suffers from many viral, fungal and bacterial diseases. Serious viral

diseases include cowpea yellow mosaic virus, cowpea aphid-borne mosaic virus, cowpea mottle virus and golden mosaic virus. Crop rotation, weeding to remove alternative hosts and resistant cultivars are necessary for integrated control.
Diseases caused by fungi and bacteria are classified according to plant parts or growth stages most adversely affected. Serious stem diseases in Africa include anthracnose (*Colletotrichum lindemuthianum*); major foliar diseases include bacterial blight (*Xanthomonas vignicola*), *Cercospora* leafspots and several rusts. Wilt (*Fusarium oxysporum*) is common in Asia. *Sphaceloma* scab and brown blotch (*Colletotrichum* spp.) are serious pod diseases.
The major diseases can be controlled by appropriate cultural practices, the use of resistant cultivars and by integrated management that implies the complementary use of different control methods. The serious mycoplasm disease is phyllody. Cowpea is attacked by many insect pests throughout its geographic range but especially in Africa. The pests include aphids, beanfly, leafhoppers, thrips, podborers, pod-sucking bugs, cowpea curculio and storage weevils. In Africa, insect pests are often responsible for yield losses, sometimes even crop failure. Serious pests in Africa are: aphids (*Aphis craccivora*), thrips, pod bugs, pod borers (*Maruca testulalis*) and storage weevils (*Callosobruchus* spp.). In Asia, serious pests seem to be aphids, leaf hoppers (*Empoasca* spp.), beanfly (*Ophiomyia phaseoli*) and storage weevils. Control measures include the use of insect-resistant cultivars in combination with cultural control methods and application of insecticides in minimal amounts. Economic losses also occur from nematodes like root-knot nematodes (*Meloidogyne* spp.) and reniform nematodes (*Rotylenchulus* spp.). Control measures include crop rotation, fallow and resistant cultivars.

Harvesting Green pods are harvested by hand when they are still immature and tender (12–15 days after flowering). Pod picking may continue for 6–8 weeks for yard-long bean. When grown as a pulse, harvesting is complicated by the prolonged and uneven ripening of many cultivars. Time of harvesting is critical as mature pods easily shatter. So they are hand-picked now and then. Sometimes plants are pulled when most of the pods are mature. For hay, the crop is cut when most of the pods are well developed. Harvesting is done by hand or mechanically.

Yield Under subsistance agriculture in Africa, average yield of seed is 100–300 kg/ha. When grown as a sole crop with good management, cowpea can yield 1000–4000 kg/ha. Largest yields have been achieved by crops late to flower and mature. Yields of dry seeds in South-East Asia are usually higher than in West Africa. Yields of catjang cowpea in India are 1000–2500 kg/ha. Yield of green pods of yard-long bean is 6–8 t/ha.

Handling after harvest After threshing, seed should be thoroughly dried to a moisture content of 14 % or less before being stored. Cowpeas are extremely susceptible to insect infestation during storage. Farmers may treat seeds with palm, groundnut or coconut oil as protection during storage. Most of the harvest is sold and consumed locally. Considerable amounts of cowpeas are processed in the United States. Yard-long beans may be quick-frozen but since the pods wilt quickly after picking, they must be processed with a minimum of delay in order to achieve an acceptable product that can compete with snap beans (*Phaseolus vulgaris*).

Genetic resources The most extensive collection of germplasm is maintained at the International Institute of Tropical Agriculture, Ibadan, Nigeria, with 10 000 accessions. However germplasm from South-East Asia is not well represented. There is a need to collect local Asian germplasm and to expand the collections of certain areas of cowpea and wild relatives for successful crop improvement. The collections of the weedy and closely related species of cowpea are limited. The gene pool that can currently be exploited by cowpea breeders comprise the cultivar groups and their landraces as well as all the wild subspecies of *Vigna unguiculata*.

Breeding Insect pests and pathogens are the principal factors limiting productivity throughout Africa, whereas both sensitivity to photoperiod and intermediate plant types are relevant in traditional farming. The strategy of IITA is now to incorporate pest and disease resistance and photoperiodic reactions into a range of plant types, suited to different cropping systems and environments. Breeding for resistance to drought and heat is necessary for the production of cowpeas in semi-arid zones. Considerable emphasis is placed on developing extra early cowpea cultivars for areas with short rainy seasons and for areas where a catch crop after rice or wheat is possible (60–90 days). Breeding work on yard-long bean is almost non-existent. The University of the Philippines, Los Baños, developed several cultivars by crossing yard-long bean and common cowpea (so-called bush sitao) for vegetable purpose. To improve the productivity of yard-long bean, breeding for resistance to pests and diseases and improved plant types are required. The cultivated and wild forms of cowpea in southern Africa are promising for improvement programmes.

Prospects The prospects for *V. unguiculata* are reasonably good through crop improvement and better management practices. Efforts to gather genetic resources and use in breeding are the keys for future improvement. Cowpea is of great value in multiple cropping systems involving relay cropping and mixed intercropping. It also has great potential as a short-duration catch crop in several Asian countries. As the demand for vegetables will grow in South-East Asia, yard-long bean may play a useful role in supplying proteins and vitamins.

Literature |**1**| Duke, J.A., 1981. Handbook of legumes of world economic importance. Plenum Press, New York and London. p. 298–306. |**2**| Kay, D.E., 1979. Food legumes. Tropical Products Institute, London. Crop and Product Digest No 3. p. 11–16, 86–114. |**3**| Lush, W.M. & Evans, L.T., 1981. Domestication and improvement of cowpea. Euphytica 30:579–587. |**4**| Maréchal, R., Mascherpa, J.-M. & Stainier, F., 1978. Etude taxonomique d'une groupe complexe d'espèces des genres Phaseolus et Vigna (Papilionaceae) sur la base de données morphologiques et polliniques, traitées par l'analyse informatique. Boissiera 28:191–196. |**5**| Rawal, K.M., 1975. Natural hybridization among wild, weedy and cultivated Vigna unguiculata (L.) Walp. Euphytica 24:699–707. |**6**| Singh, S.R. & Rachie, K.O. (Editors), 1985. Cowpea research, production and utilization. John Wiley & Sons, Chichester, Essex. 460 pp.

R.K. Pandey & E. Westphal

3 Information on minor pulses

Padbruggea dasyphylla Miq.

Synonyms *Millettia dasyphylla* (Miq.) Boerl. ex Koord.

Vernacular names Indonesia: Buhu, mata buhu, oyod munding.

Distribution Indonesia and Malay Peninsula.

Uses Seeds eaten boiled or roasted.

Observations Robust liana, 5–35 m, in forest borders and along watersides. Seeds injurious if eaten raw.

Selected sources 1, 2, 3, 8.

Shuteria hirsuta Baker

Synonyms *Pueraria anabaptista* Kurz, *Shuteria anabaptista* (Kurz) Wu.

Vernacular names Laos: sa 'thoua, tang hmeng dô. Vietnam: mang sang, cam thao núi, dâu ma.

Distribution Burma, Laos, Thailand, Vietnam, India, China.

Uses (Dry?) seeds edible.

Observations Liana, 3–4 m, in open forests, along waysides, on rocky soils.

Selected sources 6

Vigna minima (Roxb.) Ohwi & Ohashi

Synonyms *Phaseolus minimus* Roxb., *Azukia minima* (Roxb.) Ohwi.

Distribution Philippines, China, Japan, Ryukyu Islands, Taiwan.

Uses (Dry?) seeds edible.

Observations Annual herb, 1–2 m long, twining, in thickets and among coarse grasses. Three formae have been described, here f. *minima*.

Selected sources 4, 5, 7.

Sources of literature

1. Backer, C.A. & Bakhuizen van den Brink, R.C., 1963. Flora of Java. Rijksherbarium, Leyden. Noordhoff, Groningen. Vol. 1, p. 616.
2. Burkill, I.H., 1966. A dictionary of the economic products of the Malay Peninsula. 2nd ed. Kuala Lumpur. Vol. 2, p. 1648.
3. Heyne, K., 1927. De nuttige planten van Nederlandsch Indië. 2nd ed. Departement van Landbouw, Nijverheid en Handel in Nederlandsch-Indië. Batavia. Vol. 2, p. 797.
4. Maréchal, R. et al., 1978. Boissiera 28: 216–217.
5. Merrill, E.D., 1912. A flora of Manila. Manila Bureau of Printing. p. 261.
6. Nguyên Van Thuân, 1979. Flore du Cambodge, du Laos et du Viêt-Nam 17. Légumineuses–Papilionoïdées. Muséum National d'Histoire Naturelle, Paris. pp. 67–68, ill. p. 65, No 11, 12.
7. Ohashi, H., 1977. In: Flora of Taiwan. Epoch Publishing Co., Taipei, Taiwan. Vol. 3, pp. 413–415.
8. Ridley, H.N., 1922. The flora of the Malay Peninsula. The Government of the Straits Settlements. London. Vol. 1, p. 587.

4 Legumes only occasionally used as a pulse

In this list, the commodity group is given in parenthesis and the synomyms are given in the following, indented lines.

Abrus precatorius L. (medicinal and poisonous plants)
Acacia leucophloea (Roxb.) Willd. (dye and tannin-producing plants)
Adenanthera pavonina L. (timber trees)
Afzelia xylocarpa (Kurz) Craib (timber trees)
    Pahudia xylocarpa Kurz
    Pahudia cochinchinensis Pierre
Albizia acle (Blanco) Merr. (timber trees)
    Serialbizia acle (Blanco) Kosterm.
Alysicarpus rugosus (Willd.) DC. (forages)
    Hedysarum rugosum Willd.
    Alysicarpus wallichii Wight & Arn.
Bauhinia racemosa Lam. (auxiliary plants in agriculture and forestry)
    Piliostigma racemosa (Lam.) Benth.
    Bauhinia parviflora Vahl
Bauhinia vahlii Wight & Arn. (dye and tannin-producing plants)
Caesalpinia digyna Rottler (dye and tannin-producing plants)
Caesalpinia pulcherrima (L.) Swartz (ornamental plants)
Canavalia ensiformis (L.) DC. (auxiliary plants in agriculture and forestry)
Canavalia gladiata (Jacq.) DC. (vegetables)
Canavalia maritima (Aublet) Piper (auxiliary plants in agriculture and forestry)
    Canavalia obtusifolia (Lam.) DC.
    Canavalia rosea (Swartz) DC.
Cassia bicapsularis L. (ornamental plants)
Cassia floribunda Cav. (ornamental plants)
    Cassia laevigata Willd.
Cassia tora L. (medicinal and poisonous plants)
Castanospermum australe Cunn. & C. Fraser ex Hook. (timber trees)
Ceratonia siliqua L. (forages)
Clitoria ternatea L. (auxiliary plants in agriculture and forestry)
Crotalaria juncea L. (auxiliary plants in agriculture and forestry)
Crotalaria pallida Aiton (auxiliary plants in agriculture and forestry)
    Crotalaria mucronata Desv.
    Crotalaria striata DC.
Crotalaria retusa L. (auxiliary plants in agriculture and forestry)
Cyamopsis tetragonoloba (L.) Taubert (auxiliary plants in agriculture and forestry)
    Cyamopsis psoralioides DC.

Desmodium repandum (Vahl) DC. (medicinal and poisonous plants)
  Desmodium scalpe DC.
Dialium cochinchinense Pierre (timber trees)
Entada phaseoloides (L.) Merr. (medicinal and poisonous plants)
Entada pursaetha DC. (medicinal and poisonous plants)
  Entada spiralis Ridley sensu Burkill
Erythrina variegata L. (auxiliary plants in agriculture and forestry)
  Erythrina variegata L. var. *orientalis* (L.) Merr.
Indigofera cordifolia K. Heyne ex Roth (forages)
Indigofera glandulosa Wendl. (forages)
  Indigofera glandulosa Roxb. ex Willd.
  Psoralea leichhardtii F. v. Muell.
Indigofera linifolia (L.f.) Retz. (medicinal and poisonous plants)
Indigofera linaei Ali (medicinal and poisonous plants)
  Hedysarum prostratum L.
  Indigofera enneaphylla L.
  Indigofera prostrata (L.) Domin
  Indigofera dominii Eichler
Inga laurina Willd. (auxiliary plants in agriculture and forestry)
Inocarpus fagiferus (Parkinson) Fosb. (ornamental plants)
  Inocarpus edulis J.R.& G. Forster
Intsia bijuga (Colebr.) Kuntze (timber trees)
  Intsia amboinensis DC.
  Intsia retusa (Kurz) Kuntze
  Afzelia retusa Kurz
Lathyrus odoratus L. (ornamental plants)
Leucaena leucocephala (Lam.) De Wit (auxiliary plants in agriculture and forestry)
  Leucaena glauca (Willd.) Benth.
Lysidice rhodostegia Hance (ornamental plants)
Macroptilium lathyroides (L.) Urban var. *semierectum* (L.) Urban (auxiliary plants in agriculture and forestry)
  Phaseolus semierectus L.
  Phaseolus lathyroides L.
Mucuna gigantea (Willd.) DC. (medicinal and poisonous plants)
Mucuna pruriens (L.) DC. cv. group Utilis (auxiliary plants in agriculture and forestry)
  Mucuna deeringiana (Bort.) Merr.
  Mucuna aterrima (Piper & Tracy) Merr.
  Mucuna cochinchinensis (Lour.) A. Chev.
  Mucuna utilis Wall. ex Wight
  Mucuna pruriens (L.) DC. var. *utilis* (Wall. ex Wight) Baker ex Burck
Parkia intermedia Hassk. (spices and condiments)
Parkia leiophylla Kurz (vegetables)
Parkia roxburghii G. Don (medicinal and poisonous plants)
  Parkia javanica (Lam.) Merr.
Parkia speciosa Hassk. (spices and condiments)
Parkinsonia aculeata L. (ornamental plants)
Piliostigma malabaricum (Roxb.) Benth. var. *acidum* (Korth.) De Wit (timber trees)
  Bauhinia malabarica Roxb.

Pithecellobium bubalinum Benth. (timber trees)
Pithecellobium dulce (Roxb.) Benth. (dye and tannin-producing plants)
  Mimosa dulcis Roxb.
  Inga dulcis (Roxb.) Willd.
Pithecellobium fagifolium Blume ex Miq. (spices and condiments)
  Zygia fagifolia (Blume ex Miq.) Kosterm.
Pithecellobium jiringa (Jack) Prain ex King (spices and condiments)
  Pithecellobium lobatum Benth.
Pithecellobium unguis-cati (L.) Benth. (ornamental plants)
Pongamia pinnata (L.) Pierre (medicinal and poisonous plants)
  Cytisus pinnatus L.
  Pongamia mitis (L.) Merr.
Psophocarpus scandens (Endl.) Verdc. (auxiliary plants in agriculture and forestry)
  Psophocarpus longepedunculatus Hassk.
  Psophocarpus palustris Desv. sensu Heyne, Burkill, Backer
Psophocarpus tetragonolobus (L.) DC. (vegetables)
Pterocarpus marsupium Roxb. (timber trees)
Saraca dives Pierre (forages)
Schizolobium parahybum (Vell.) Blake (ornamental plants)
  Schizolobium excelsum Vog.
Sesbania bispinosa (Jacq.) W.F. Wight (fibre plants)
  Sesbania aculeata Pers.
Sesbania grandiflora (L.) Poiret (auxiliary plants in agriculture and forestry)
Tamarindus indica L. (spices and condiments)
Tephrosia purpurea (L.) Pers. (auxiliary plants in agriculture and forestry)
Trigonella foenum-graecum L. (spices and condiments)
Vigna marina (Burman) Merrill (auxiliary plants in agriculture and forestry)
  Phaseolus marinus Burman
  Vigna anomala Walpers
  Vigna retusa Walpers
  Vigna oblonga Bentham
  Vigna lutea (Swartz) A. Gray
  Phaseolus obovatus Grah. ex Gagnep.
Vigna pilosa (Roxb.) Baker (auxiliary plants in agriculture and forestry)
Vigna sublobata (Roxb.) Babu et Sharma (auxiliary plants in agriculture and forestry)
  Phaseolus sublobatus Roxb.
Vigna trilobata (L.) Verdc. (forages)
  Dolichos trilobatus L.
  Phaseolus trilobatus (L.) Schreb.
Vigna vexillata (L.) A. Rich. (auxiliary plants in agriculture and forestry)
Xylia xylocarpa (Roxb.) Taub. (timber trees)
  Xylia kerrii Craib & Hutchinson

Literature

Frequently cited references, such as general reference works, the recent volumes on advances in legume science and systematics, are combined here, and omitted from the reference lists provided with each species.

Aykroyd, W.R. & Doughty, J., 1982. Legumes in human nutrition. 2nd ed. FAO Nutritional Studies. FAO, Rome. 138 pp.

Backer, C.A. & Bakhuizen van den Brink, R.C., 1963. Flora of Java I. Leguminosae. Noordhoff, Groningen. pp. 523–645.

Bressani, R. & Elias, L.G., 1980. Nutritional value of legume crops for humans and animals. In: Summerfield, R.J. & Bunting, A.H. (Editors): Advances in legume science. Royal Botanic Gardens, Kew. pp. 135–155.

Brown, W.H., 1951–1957. Useful plants of the Philippines. Reprint of the 1941–1943 ed. Department of Agriculture and National Resources. Technical Bulletin 10. Manila, Bureau of Printing. Vol. 1 (1951), 590 pp. Vol. 2 (1954), 513 pp. Vol. 3 (1957), 507 pp.

Burkill, I.H., 1966. A dictionary of the economic products of the Malay Peninsula. 2nd ed. Ministry of Agriculture and Co-operatives, Kuala Lumpur. Vol. 1 (A–H), pp. 1–1240. Vol. 2 (I–Z), pp. 1241–2444.

Cobley, L.S. & Steele, W.M., 1976. An introduction to the botany of tropical crops. 2nd ed. Longman, London. pp. 71–110.

Corby, D.H.L., 1981. The sytematic value of leguminous root nodules. In: Polhill, R.M. & Raven, P.H. (Editors): Advances in legume systematics. Part 1. Royal Botanic Gardens, Kew. pp. 657–669.

Crevost, Ch., et al., 1917–1941. Catalogue des produits de l'Indochine. Gouvernement Général de l'Indochine, Hanoi. Vol. 1 (1917), 486 pp. Vol. 2 (1919–1921), 297 pp. Vol. 3 (1922–1924), 177 pp. Vol. 4 (1927), 197 pp. Vol. 5.1 (1928), 338 pp. Vol. 5.2 (1935), 322 pp. Vol. 6 (1941), 125 pp.

Duke, J.A., 1981. Handbook of legumes of world economic importance. Plenum Press, New York & London. 345 pp.

FAO, 1985. Production yearbook. FAO, Rome.

FAO, 1988. Data for selected commodities in S.E. Asia. Food and Agriculture Organization of the United Nations, Rome. (Unpublished database)

Fehr, W.R. & Hadley, H.H. (Editors), 1980. Hybridization of crop plants. American Society of Agronomy & Crop Science. Society of America Inc. Madison, Wisconsin. 765 pp.

Harlan, J.R. & De Wet, J.M.J., 1971. Toward a rational classification of cultivated plants. Taxon 20(4): 507–571.

Herridge, D.F., Sudin, M.N., Pate, J.S. & Peoples, M.B., 1988. Translocation and utilization of nitrogen by pulses. In: Summerfield, R.J. (Editor): World crops – Cool season food legumes. Kluwer Academic Publishers, Dordrecht, Boston & London. pp. 793–800.

Heyne, K., 1927. De nuttige planten van Nederlandsch Indië. 2nd ed. 3 volumes. Departement van Landbouw, Nijverheid en Handel in Nederlandsch-Indië. 1953 pp.

IBPGR, 1980. Directory of germplasm collections I. Food legumes. Compiled by: Ayad, G. & Murthi Anishetty, N. IBPGR, Rome. 22 pp.

IBPGR, 1984. Directory of germplasm resources. Forages (grasses, legumes, etc.). Compiled by: Ellis Davies, W. & McLean, B.T. IBPGR, Rome. 43 pp.

IBPGR, 1985. Directory of germplasm collections I.II. Food legumes (soyabean). Compiled by: Juvik, G.A., Bernard, R.L. & Kauffman, H.E. IBPGR, Rome. 53 pp.

IBPGR, 1987. A catalogue of Chinese crop germplasm for exchange. IBPGR, Rome. 97 pp.

ICRISAT, 1984. Grain legumes in Asia. Summary Proceedings of the Consultative Group Meeting for Asian Regional Research on Grain Legumes (Groundnut, Chickpea, Pigeonpea). ICRISAT Centre, 11–15 December 1983. ICRISAT, Patancheru, A.P., India. 98 pp.

ICRISAT, 1987. Asian grain legumes network. AGLN Cooperators Report. Legumes program. ICRISAT, Patancheru, A.P., India. 38 pp.

Kay, D.E., 1979. Food legumes. TPI Crop and Product Digest No 3. Tropical Products Institute, London. 435 pp.

Lackey, J.A., 1981. Phaseoleae. In: Polhill, R.M. & Raven, P.H. (Editors): Advances in legume systematics. Part 1. Royal Botanic Gardens, Kew. pp. 301–327.

McComb, J.A., 1975. Is intergenetic hybridization possible in the Leguminosae? Euphytica 24:497–502.

NAS, 1979. Tropical legumes, resources for the future. National Academy of Sciences, Washington, D.C. 331 pp.

Nguyên Van Thuân, 1979. Légumineuses-Papilionoïdées, Phaséolées. Flore du Cambodge, du Laos et du Viêt-Nam 17. Muséum National d'Histoire Naturelle, Paris. 217 pp.

Nguyên Van Thuân, Dy Phon, P., Niyomdham, C. & Vidal, Y., 1987. Légumineuses-Papilionoïdées. Flore du Cambodge, du Laos et du Viêtnam 23. Muséum National d'Histoire Naturelle, Paris. 258 pp.

Norton, G., Bliss, F.A. & Bressani, R., 1985. Biochemical and nutritional attributes of grain legumes. In: Summerfield, R.J. & Roberts, E.H. (Editors): Grain legume crops. Collins, London. pp. 73–114.

Pate, J.S., Atkins, C.A., Peoples, M.B. & Herridge, D.F., 1988. Partitioning of carbon and nitrogen in the nodulated grain legume: principles, processes and regulation. In: Summerfield, R.J. (Editor): World crops – Cool season food legumes. Kluwer Academic Publishers, Dordrecht, Boston & London. pp. 751–765.

Pate, J.S. & Minchin, F.R., 1980. Comparative studies of carbon and nitrogen nutrition of selected grain legumes. In: Summerfield, R.J. & Bunting, A.H. (Editors): Advances in legume science. Royal Botanic Gardens, Kew. pp. 105–114.

Polhill, R.M. & Van der Maesen, L.J.G., 1985. Taxonomy of grain legumes. In: Summerfield, R.J. & Roberts, E.H. (Editors): Grain legume crops. Collins, London. pp. 3–36.

Polhill, R.M. & Raven, P.H., 1981. Advances in legume systematics. Parts 1 & 2. Royal Botanic Gardens, Kew. 1049 pp.

Purseglove, J.W., 1968. Tropical crops. Dicotyledons 1. Longman, London. pp. 199–332.

Rachie, K.O. & Roberts, L.M., 1974. Grain legumes of the lowland tropics. Advances in Agronomy 26: 1–132.

Simmonds, N.W. (Editor), 1976. Evolution of crop plants. Longman, London & New York. 339 pp.

Smartt, J., 1976. Tropical pulses. Longman, London. 348 pp.

Sprent, J.I. & Minchin, F.R., 1985. Rhizobium, nodulation and nitrogen fixation. In: Summerfield, R.J. & Roberts, E.H. (Editors): Grain legume crops. Collins, London. pp. 115–144.

Stirton, C.H. (Editor), 1987. Advances in legume systematics. Part 3. Royal Botanic Gardens, Kew. 466 pp.

Summerfield, R.J. & Roberts, E.H. (Editors), 1985. Grain legume crops. Collins, London. 859 pp.

Summerfield, R.J. & Bunting, A.H. (Editors), 1980. Advances in legume science. Royal Botanic Gardens, Kew. 667 pp.

Summerfield, R.J. & Wien, H.C., 1980. Effects of photoperiod and air temperature on growth and yield of economic legumes. In: Summerfield & Bunting (Editors): Advances in legume science. Royal Botanic Gardens, Kew. pp. 17–36.

Tucker, S.C., 1987. Floral initiation and development in legumes. In: Stirton (Editor): Advances in legume systematics. Part 3. Royal Botanic Gardens, Kew. pp. 183–232.

Van der Maesen, L.J.G., 1986. Cajanus DC. and Atylosia W. & A. (Leguminosae). Agricultural University Wageningen Papers 85–4: 21–22.

Van der Maesen, L.J.G., 1989 (in press). Genetic resources of tropical legumes. In: Proceedings Symposium on Systematic Botany – A key science for tropical research and documentation, Stockholm/Uppsala. Sept. 1987.

Verdcourt, B., 1970. Studies in the Leguminosae-Papilionoideae for the 'Flora of Tropical East Africa': IV. Kew Bulletin 24: 507–569.

Verdcourt, B. 1979. A manual of New Guinea legumes. Botany Bulletin No 11. Office of Forests, Division of Botany, Lae, Papua New Guinea. 645 pp.

Wallis, E.S. & Byth, D.E. (Editors), 1987. Food legume improvement for Asian farming systems. Proceedings International Workshop held at Khon Kaen, Thailand, 1–5 September 1986. ACIAR Proceedings No 18. 341 pp.

Webb, C. & Hawtin, G. (Editors), 1981. Lentils. C.A.B./ICARDA, Farnham Royal. 216 pp.

Westphal, E., 1974. Pulses in Ethiopia, their taxonomy and agricultural significance. Agricultural Research Reports 815. Pudoc, Wageningen. 261 pp.

Zwartz, J.A., & Hautvast, J.G.A.J. 1979. Food supplies, nutrition and plant breeding. In: Sneep, J. & Hendriksen, A.J.T. (Editors): Plant breeding perspectives – Centennial publication of Koninklijk Kweekbedrijf en Zaadhandel D.J. van der Have 1879–1979. Pudoc, Wageningen. pp. 1–46.

Glossary

actinomorphic: radially symmetrical; applied to flowers that can be bi-sected in more than one vertical plane

acuminate: ending in a narrowed, tapering point with concave sides

acute: sharp; ending in a point with straight or slightly convex sides

adaxial: the side or face next to the axis

adnate: united with another part; with unlike parts fused, e.g. ovary and calyx tube

adventitious: not in the usual place, e.g. roots on stems, or buds produced elsewhere than in the axils of leaves or the extremities of stems

ala, pl. alae: wing(s), the lateral petal(s) of a papilionaceous flower

alternate: leaves, etc., inserted at different levels along the stem, as distinct from opposite or whorled

androecium: the male element; the stamens as a unit of the flower

aneuploid: with other than the exact multiple of the haploid chromosome complement

annual: a plant that completes its life cycle in one year

anther: the part of the stamen containing the pollen

anthesis: the time the flower is expanded, or, more strictly, the time when pollination takes place

apiculate: ending abruptly in a short point

appressed: lying closely and flatly pressed against

aril: an expansion of the funicle enveloping the seed, arising from the placenta; sometimes occurring as a pulpy covering

attenuate: gradually tapering

axil: the upper angle between the leaf and the stem

axillary: arising from the axil

axis: the main or central line of development of any plant or organ

beak: a long, prominent and substantial point, applied particularly to prolongations of fruits

bifid: cleft into two parts at the tip

bisexual: having both sexes present and functional in the same flower

blade: the expanded part of a leaf or petal

bract: a reduced leaf subtending a flower or flower stalk, or a part of an inflorescence

bracteole: a secondary bract on the pedicel or close under the flower

bunch: cluster, growing together

bush: a low thick shrub without a distinct trunk

calyx: the outer envelope of the flower, consisting of sepals, free or united

campanulate: bell-shaped

capitate: headed, like the head of a pin in some stigmas, or collected into compact

headlike clusters as in some inflorescences
carina: keel, the two inner united petals of a papilionaceous flower
carpel: one of the foliar units of a compound pistil or ovary; a simple pistil has only one carpel
caruncle: an outgrowth of a seed near the hilum
cataphylls: scale leaves (early leaf forms) in e.g. hypogeal germinating seedlings, which appear before the eophylls
ciliate: with a fringe of hairs along the edge
cleistogamous: when self-pollination occurs within the unopened flower
compound: of two or more similar parts in one organ, as in a compound leaf or compound fruit
concave: hollow
connate: united or joined
cordate: heart-shaped, as seen at the base of a deeply notched leaf, etc.
coriaceous: of leathery texture
corolla: the inner envelope of the flower of free or united petals
cotyledon: seed-leaf; dicotyledons have two cotyledons in their embryos and monocotyledons have one
cross-pollination: placing or depositing the pollen from one flower on the stigma of a flower of another plant
cultivar (cv., cvs): an agricultural or horticultural variety that has originated and persisted under cultivation, as distinct from a botanical variety; a cultivar name should always be written with an initial capital letter and given single quotation marks, e.g., banana 'Gros Michel'
cuneate: wedge-shaped; triangular, with the narrow end at the point of attachment, as the bases of leaves or petals
deciduous: shedding or prone to shedding, applied to leaves, petals, etc.
deflexed (reflexed): abruptly recurved; bent downwards or backwards
dehiscent: opening spontaneously when ripe, e.g., capsules, anthers
deltoid: shaped like an equal-sided triangle
determinate: when the terminal or central flower of an inflorescence opens first and the prolongation of the axis is arrested; for pulses also used to indicate bush-shaped plants with short duration flowering in one plane
diadelphous: in two bundles
dimorphic: of two forms, as may occur with branches, etc.
dioecious: with unisexual flowers and with the staminate and pistillate flowers on different plants
diploid: with two sets (genomes) of chromosomes, as occurs in somatic or body cells; usually written $2n$: having twice the basic chromosome number of the haploid germ cells
distichous: regularly arranged in two opposite rows on either side of the stem
downy: covered with very short and weak soft hairs
elliptic: oval in outline but widest about the middle
endemic: confined to a region or country and not native anywhere else
endocarp: the innermost layer of the pericarp or fruit wall
entire: an even margin without teeth, lobes, etc.
eophylls: seedling leaves, as distinct from adult leaves called metaphylls
epicotyl: the young stem above the cotyledons
epigeal: above ground; in epigeal germination cotyledons appear above ground

exocarp: the outer layer of the pericarp or fruit wall
exserted: projecting beyond, as stamens from a perianth
extrorse: an anther which dehisces outwardly towards the perianth
falcate: sickle-shaped
fascicle: a cluster of flowers, leaves, etc., arising from the same point
fertilization: union of the gametes (egg and sperm) to form a zygote
filament: thread; the stalk supporting the anther
filiform: slender; threadlike
flexuose: zigzag; bent alternately in opposite directions
foliolate (2-, 3-, 4- etc.): with 2-, 3-, 4- leaflets
funicle (funiculus): the little cord which attaches the ovule or seed to the placenta
genome: a set of chromosomes as contained within the gamete and corresponding to the haploid chromosome number of the species
genotype: the genetic makeup of an organism comprising the sum total of its genes, both dominant and recessive; a group of organisms with the same genetic makeup
gibbous: more convex in one place than another
glabrescent: becoming glabrous or nearly so
glabrous: devoid of hairs
glandular: having or bearing secreting organs or glands
globose: spherical or nearly so
gynoecium: the female part or pistil of a flower, consisting, when complete, of one or more ovaries with their styles and stigmas
gynophore: a stalk supporting the gynoecium formed by elongation of the receptacle
harvest index: the total harvested (used product) in relation to the total biomass of the crop
herb: any vascular plant which is not woody
herbaceous: not woody
hilum: the scar left on a seed indicating its point of attachment
husk: the outer covering of some fruits
hybrid: the first generation offspring of a cross between two individuals differing in one or more genes
hybridization: the crossing of individuals of unlike genetic constitution
hypanthium: the cup-like receptacle usually derived from the fusion of the floral envelopes and androecium on which are seemingly borne the calyx, corolla and stamens
hypocotyl: the young stem below the cotyledons
hypogeal: below ground; in hypogeal germination the cotyledons remain below ground within the testa
imparipinnate: pinnate with an odd terminal leaflet
incompatibility: failure to obtain fertilization and seed formation after self-pollination, or within or between clones
indehiscent: not opening when ripe
indeterminate: an inflorescence in which the terminal flowers are the last to open, so that the floral axis may be prolonged indefinitely by a terminal bud; in pulses also used to indicate plants with climbing stems with long-duration flowering

indigenous: native to a particular area or region
inflorescence: the arrangement and mode of development of the flowers on the floral axis
intercalary: growth, not apical but between the apex and the base
keel: see carina
lamina: see blade
lanceolate: lance-shaped; much longer than broad, being widest at the base and tapering to the apex
lateral: on or at the side
leaflet: one part of a compound leaf
linear: long and narrow with parallel sides
lobed: of leaves: divided, but not into separate leaflets
malesia: the biogeographical region from NW Sumatra to New Britain, embracing the Malay Archipelago
meristem: undifferentiated tissue of the growing point whose cells are capable of dividing and developing into various organs and tissues
mesocarp: the middle layer of the pericarp or fruit wall which is often fleshy or succulent
metaphylls: adult leaves
monadelphous: of stamens which are united into one group by their filaments
monoecious: with unisexual flowers but borne on the same plant
mycorrhiza: a symbiotic association of roots with a fungus which may form a layer outside the root (ectotrophic) or within the outer tissues (endotrophic)
node: the point on the stem or branch at which a leaf or branch is borne
nodule: a small knot or rounded body, often in roots of leguminous plants, where bacteria of the genus *Rhizobium* are active
oblanceolate: reverse of lanceolate
oblong: longer than broad with the sides parallel or almost so
obovate: reverse of ovate
obtuse: blunt or rounded at the end
opposite: of leaves and branches when two are borne at the same node on opposite sides of the stem
orbicular: flat with a more or less circular outline
oval: see ovate
ovary: that part of the pistil, usually the enlarged base, which contains the ovules and eventually becomes the fruit
ovate: egg-shaped; a flat surface which is scarcely twice as long as broad with the widest portion below the middle
ovoid: a solid object which is egg-shaped (ovate) in section
ovule: the immature seeds in the ovary before fertilization
panicle: an indeterminate branched racemose inflorescence
paniculate: resembling a panicle
papilionaceous flower: butterfly-like, pea-like flower, with standard, wings and keel
paripinnate: a pinnate leaf without the odd terminal leaflet
pedicel: stalk of each individual flower of an inflorescence
peduncle: the stalk of an inflorescence or partial inflorescence
perennial: living from year to year and usually flowering each year
perianth: the floral leaves as a whole, including both speals and petals if both are present

pericarp: the wall of the ripened ovary or fruit wall of which the layers may be fused into one, or be more or less divisible into exocarp, mesocarp and endocarp
persistent: remaining attached; not falling off
petal: a member of the inner series of perianth segments which are often brightly coloured
petaloid: petal-like
petiole: the stalk of a leaf
petiolule: the stalk of a leaflet
phyllody: transformation of flower parts into leaves
phyllotaxy: the arrangement of leaves or floral parts on their axis
pinna, pl. pinnae: a primary division or leaflet of a pinnate leaf
pinnate: a compound leaf with the leaflets arranged along each side of a common rachis
pistil: the female part of a flower (gynoecium) of one or more carpels, consisting, when complete, of ovary(s), style(s) and stigma(s)
pod: a general term for a dry dehiscent fruit
pollen: spores or grains borne by the anthers containing the male element (gametophyte)
polyploid: an organism with more than two sets (genomes) of chromosomes in its somatic cells
primordium: a group of undifferentiated meristematic cells, usually of a growing point, capable of differentiating into various kinds of organs or tissues
prostrate: lying flat on the ground
pseudoraceme: raceme-like inflorescence but not a true raceme
pubescent: covered with soft short hairs
pulvinus: a minute gland or swollen petiole base
punctate: marked with dots or translucent glands
qualitative short-day plant: to flower, the plant needs short days (often with quantitative response); if the daylength surpasses a certain value (the critical daylength) the plant does not flower
quantitative short-day plant: plant flowers sooner under short-day conditions, but short days are not absolutely necessary to flower
raceme: an unbranched elongated indeterminate inflorescence with stalked flowers opening from the base upwards
racemose: raceme-like
rachis: the principal axis or an inflorescence or a compound leaf
recumbent: lying down
recurved: bent or curved downward or backward
reniform: kidney-shaped
reticulate: netted, as when the smallest veins of a leaf are connected together like the meshes of a net
rhizome: an underground stem which is distinguished from a root by the presence of nodes, buds, and leaves or scales
rhomboid: quadrangular, with the lateral angles obtuse
scale: reduced leaf, usually sessile, thin and dry, and seldom green
seed: the reproductive unit formed from a fertilized ovule, consisting of embryo and seed-coat, and, in some cases, also endosperm
self-fertile: capable of fertilization and setting seed after self-pollination

self-pollination: pollination with pollen from the same flower or from other flowers of the same plant
self-sterile: failure to complete fertilization and obtain seed after self-pollination
sepal: a member of the outer series of perianth segments
shrub: a woody plant with branches from the base and not reaching any great size
simple: not compound, as in leaves with a single blade
stamen: one of the male reproductive organs of a flower; a unit of the androecium
staminode: an abortive or rudimentary stamen without a perfect anther
standard: the upper and outermost petal of a papilionaceous flower
sterile: failing to complete fertilization and produce seed as a result of defective pollen or ovules; not producing seed capable of germination; lacking functional sexual organs
stigma: the portion of the pistil which receives the pollen
stipe: the stalk supporting a carpel or gynoecium
stipel: small secondary stipule at the base of a leaflet
stipule: a scale-like or leaf-like appendage at the base of a leaf petiole
straggling: extremely divergent, spreading very far apart
strain: a group of individuals of a common origin, usually a more narrowly defined group than a cultivar
strophiole: see caruncle
style: the part of the pistil connecting the ovary with the stigma
superior: of an ovary with the perianth inserted below or around its base, the ovary being attached at its base only
suture: the line of junction of two carpels; the line or mark of splitting open
taproot: the primary descending root, forming a direct continuation of the radicle
tendril: a thread-like climbing organ formed from the whole or part of a stem, leaf or petiole
tepal: a segment of a perianth, sepal or petal
terminal: borne at the end or apex
testa: the outer coat of the seed
tetraploid: having four times ($4n$) the basic number of chromosomes or twice the diploid number ($2n$)
trifoliate: three-leaved
tubercle: a small tuberlike excrescence
tuberculate: covered with warty protuberances
twining: winding spirally
unisexual: of one sex, having stamens or pistils only
variety: botanical variety which is a subdivision of a species; an agricultural or horticultural variety is referred to as a cultivar
vernalization: the treatment of seeds or bulbs before planting to hasten flowering
vexillum: see standard
viability: ability to live, grow and develop
viny: trailing or climbing
wedge-shaped: see cuneate
wing: see ala
zygomorphic: irregular flowers divisible into equal halves in one plane only

Acronyms of organizations

- AVRDC: Asian Vegetable Research and Development Center (Shanhua, Tainan, Taiwan).
- BORIF: Bogor Research Institute for Food Crops (Bogor, Indonesia).
- CAAS: Chinese Academy of Agricultural Sciences (Wuhan, Hubei, China).
- CIAT: Centro Internacional de Agricultura Tropical (Cali, Colombia).
- CRIFC: Central Research Institute for Food Crops (Bogor, Indonesia).
- CSIRO: Commonwealth Scientific and Industrial Research Organization (Canberra, Australia).
- FAO: Food and Agriculture Organization of the United Nations (Rome, Italy).
- FCRI: Food Crops Research Institute (Bangkok, Thailand).
- IBPGR: International Board for Plant Genetic Resources (Rome, Italy).
- ICARDA: International Center for Agricultural Research in the Dry Areas (Aleppo, Syria).
- ICRISAT: International Crops Research Institute for the Semi-Arid Tropics (Hyderabad, India).
- IITA: International Institute of Tropical Agriculture (Ibadan, Nigeria).
- IPB: Institute of Plant Breeding (University of the Philippines, Los Baños, the Philippines).
- INTSOY: International Soybean Program (University of Illinois, Urbana-Champaign, United States).
- MARIF: Malang Research Institute for Food Crops (Malang, Indonesia).
- NBPGR: National Bureau of Plant Genetic Resources (Delhi, India).
- NPGRL: National Plant Genetic Resources Laboratory (Los Baños, the Philippines).
- USDA: United States Department of Agriculture (Washington D.C., United States).

Acknowledgments

Our thanks are due to

- the Board of the Wageningen Agricultural University, in particular Professor Dr C.C. Oosterlee, Rector Magnificus until September 1988, for initiating and supporting the Prosea project;
- the Netherlands Ministry of Education and Science, in particular the Director-General Dr E. van Spiegel, for financial support to start the project;
- the Netherlands Ministry of Agriculture and Fisheries, for financial and other support;
- the Netherlands Ministry of Foreign Affairs, Directorate General of International Cooperation for financial support to internationalize the project;
- the Departments of Tropical Crop Science and Plant Taxonomy of the Wageningen Agricultural University, for providing facilities for the Dutch staff of Prosea;
- the Board of the trust 'Landbouw Export Bureau 1916/1918', for granting a subsidy for the publications of the project;
- the Centre for Agricultural Publishing and Documentation (Pudoc), Wageningen, in particular Drs J.M. Schippers, Director, for his support, and his staff for their cooperation and documentation facilities;
- various experts for providing vernacular names of pulses of the countries. Indonesia Dr Soedarsono Riswan of the Centre for Research and Development in Biology, Bogor; Thailand Mr Chamlong Phengklai, Director Forest Herbarium of the Royal Forest Department, Bangkok; Cambodia, Laos and Vietnam Dr J.E. Vidal, Mr Pham Hoang Ho, Mrs Dy Phon of the Laboratoire de Phanérogamie of the Muséum National d'Histoire Naturelle, Paris.

Sources of illustrations

Illustrations are redrawn and adapted from

- Ali, S.I., 1977. Flora of West Pakistan 100: *Vigna aconitifolia* (habit).
- Brown, W.H., 1954. Useful plants of the Philippines. Department of Agriculture and Natural Resources. Technical Bulletin 10. Manila Bureau of Printing (Reprint 1941–1943 edition). Vol. 2, p. 149, Fig. 66: *Phaseolus vulgaris* (habit); p. 141, Fig. 60: *Vigna umbellata* (habit).
- Department of Plant Taxonomy, Wageningen Agricultural University, the Netherlands: *Arachis hypogaea* (seeds); *Cajanus cajan*; *Cicer arietinum*; *Glycine max* (seeds); *Lablab purpureus*; *Lathyrus sativus*; *Lens culinaris*; *Macrotyloma uniflorum*; *Phaseolus acutifolius*; *Phaseolus coccineus*; *Phaseolus vulgaris* (inflorescence, seeds); *Pisum sativum*; *Vicia faba*; *Vigna aconitifolia* (fruiting branch, seeds); *Vigna angularis* (flower, seed); *Vigna radiata*; *Vigna subterranea* (seed); *Vigna umbellata* (seeds); *Vigna unguiculata*.
- Hui-lin Li et al. (Editors), 1977. Flora of Taiwan. Epoch Publishing Co., Taipei, Taiwan. Vol. 3, p. 296, Pl. 593: *Glycine max* (habit).
- Milne-Redhead, E. & Polhill, R.M. (Editors), 1971. Flora of Tropical East Africa. Leguminosae-Papilionoideae 2. Crown Agents for Oversea Governments and Administrations. p. 616, Fig. 95: *Phaseolus lunatus* (habit, flower, seeds); p. 667, Fig. 100: *Vigna subterranea* (habit, flower, fruits).
- Nguyên Van Thuân, 1979. Flore du Cambodge, du Laos et du Vietnam 17. Muséum National d'Histoire Naturelle, Paris. p. 195, Pl. 31: *Vigna angularis* (fruiting branch); p. 183, Pl. 29: *Vigna dalzelliana* (flowering and fruiting branch); p. 195, Pl. 31: *Vigna mungo* (fruiting branch, flower, seed).
- Queensland Department of Primary Industries, Botany Branch, Indooroopilly, Australia: *Arachis hypogaea* (habit, flower, fruit).

Index of scientific plant names

Page numbers printed in bold refer to main treatment.

Index of vernacular plant names

Page numbers printed in bold refer to main treatment.

The Prosea Foundation
(Plant Resources of South-East Asia)

Name, location, legal status, and structure

- Prosea is a Foundation under Indonesian law, with an international charter, domiciled in Bogor. It is an autonomous, non-profit, international agency, governed by a Board of Trustees. It seeks linkages with existing regional and international organizations;
- Prosea is an international programme focusing on the documentation of information on plant resources of South-East Asia;
- Prosea consists of a Network Office at Bogor (Indonesia) coordinating 5 Country Offices in South-East Asia, and a Publication Office in Wageningen (the Netherlands).

Participating Institutions

- Forest Research Institute of Malaysia (FRIM), Karung Berkunci 201, Jalan FRI Kepong, 52109 Kuala Lumpur, Malaysia
- Indonesian Institute of Sciences (LIPI), Widya Graha, Jalan Gatot Subroto 10, Jakarta 12710, Indonesia
- Institute of Ecology and Biological Resources, National Centre for Scientific Research (IEBR-NCSR), Nghia Do, Tu Liem, Hanoi, Vietnam
- Papua New Guinea University of Technology (UNITECH), Private Mail Bag, Lae, Papua New Guinea
- Philippine Council for Agriculture, Forestry and Natural Resources Research & Development (PCARRD), Los Banos, Laguna, the Philippines
- Thailand Institute of Scientific and Technological Research (TISTR), 196 Phahonyothin Road, Bang Khen, Bangkok 10900, Thailand
- Wageningen Agricultural University (WAU), Costerweg 50, 6701 BH Wageningen, the Netherlands

Objectives

- to document and make available the existing wealth of information on the plant resources of South-East Asia for education, extension work, research and industry;
- to make operational a computerized data bank on the plant resources of South-East Asia;
- to publish the results in the form of an illustrated, multi-volume handbook in English;
- to promote the dissemination of the information gathered.

Target groups

- those professionally concerned with plant resources in South-East Asia and working in education, extension work, research and commercial production (direct users);
- those in South-East Asia depending directly on plant resources, obtaining relevant information through extension (indirect users).

Activities

- the establishment and operation of data bases;
- the publication of books;
- the sponsorship, support and organization of training courses;
- research into topics relevant to Prosea's purpose;
- the publication and disseminaton of reports and the research results.

Implementation

The programme period has been tentatively divided into 3 phases:
- preliminary phase (1985–1986): publication of 'Plant Resources of South-East Asia, Proposal for a Handbook' (1986);
- preparatory phase (1987–1990): establishing cooperation with South-East Asia through internationalization, documentation, consultation and publication; reaching agreement on the scientific, organizational and financial structure of Prosea;
- implementation phase (1991–1995): compiling, editing and publishing of the handbook; making operational the computerized data bank with the texts and additional information; promoting the dissemination of the information obtained.

Documentation

A documentation system has been developed for information storage and retrieval called SAPRIS (South-East Asian Plant Resources Information System). It consists of 6 data bases:
- BASELIST: primarily a checklist of more than 6200 plant species;
- CATALOG: references to secondary literature;
- PREPHASE: references to literature from South-East Asia;
- ORGANYM: references to institutions and their research activities;
- PERSONYM: references to specialists;
- TEXTFILE: all PROSEA publications and additional information.

Publication

The following publications have been issued so far:
- *Basic list of species and commodity grouping*, Version 1 (1989);
- *A selection*, dealing with 86 plant resources, being a cross-section of the commodity groups (1989);
- *Pulses (PROSEA 1)* (1989); low price edition 1990;

- *Edible fruits and nuts (PROSEA 2)* (1991);
- *Dye and tannin-producing plants (PROSEA 3)* (1991);
- *Proceedings of the first PROSEA International Symposium* (1989);
- *Bibliography 1: Pulses* (Edition 1) (1990).

The following publications are foreseen for the period 1991–1992:

- *Basic list of species and commodity grouping*, Final Version.
- *Forages (PROSEA 4)*;
- *Timber trees. Major commercial timbers (PROSEA 5(1))*;
- *Rattans (PROSEA 6)*;
- *Bamboos (PROSEA 7)*;
- *Vegetables (PROSEA 8)*.

In brief, Prosea is

- an international programme, focused on plant resources of South-East Asia;
- interdisciplinary, covering the fields of agriculture, forestry, horticulture and botany;
- a research programme, making knowledge available for education and extension;
- ecologically focused on promoting plant resources for sustainable tropical land-use systems;
- committed to conservation of biodiversity;
- committed to rural development through diversification of resources and application of farmers' knowledge.

Prosea Network Office

Centre for Research & Development in Biology
Jalan Raya Juanda 22–24
P.O. Box 234
Bogor 16122, Indonesia
tel.: (0251) 322859
telex: c/o 48304 BPPHBG
fax: c/o (62) (251) 326425

Prosea Publication Office

Wageningen Agricultural University
P.O. Box 341
6700 AH Wageningen, the Netherlands
tel.: (08370) 84587
telex: 45917 BURLU
fax: (31) (8370) 84731

PROSEA – Plant Resources of South-East Asia

*The handbook**

- No 1. Pulses. L.J.G. van der Maesen and Sadikin Somaatmadja (Editors). Pudoc, Wageningen. 1989 / Prosea, Bogor. 1992 / ESCAP CGPRT Centre, Bogor. 1990. (out of print).
- No 2. Edible fruits and nuts. E.W.M. Verheij and R.E. Coronel (Editors). Pudoc, Wageningen. 1991 / Prosea, Bogor. 1992.
- No 3. Dye and tannin-producing plants. R.H.M.J. Lemmens and N. Wulijarni-Soetjipto (Editors). Pudoc, Wageningen. 1991 / Prosea, Bogor. 1992.

Bibliographies

- Bibliography 1: Pulses. Edition 1. N. Wulijarni-Soetjipto and J.S. Siemonsma (Editors). PROSEA Project, Bogor, Indonesia. 1990.
- Bibliography 3: Dye and tannin-producing plants. Edition 1. N. Wulijarni-Soetjipto and J.S. Siemonsma (Editors). PROSEA Foundation, Bogor, Indonesia/Pudoc, Wageningen, the Netherlands. 1991.

Miscellaneous

- A selection. E. Westphal and P.C.M. Jansen (Editors). Pudoc, Wageningen. 1989.
- Basic list of species and commodity grouping. Version 1. R.H.M.J. Lemmens, P.C.M. Jansen, J.S. Siemonsma, F.M. Stavast (Editors). Prosea Project, Wageningen, the Netherlands. 1989. (distributed by Pudoc, Wageningen, out of print).
- Basic list of species and commodity grouping. Final version. P.C.M. Jansen, R.H.M.J. Lemmens, L.P.A. Oyen, J.S. Siemonsma, F.M. Stavast and J.L.C.H. van Valkenburg (Editors). Pudoc, Wageningen. 1991.
- Proceedings of the First PROSEA International Symposium, May 22–25, 1989, Jakarta, Indonesia. J.S. Siemonsma and N. Wulijarni-Soetjipto (Editors). Pudoc, Wageningen. 1989. (out of print).

* The handbook in blue cover (hardbound) is distributed by Pudoc, the low-cost edition in green cover (paperback) by Prosea only in the developing countries of South-East Asia and the Pacific.

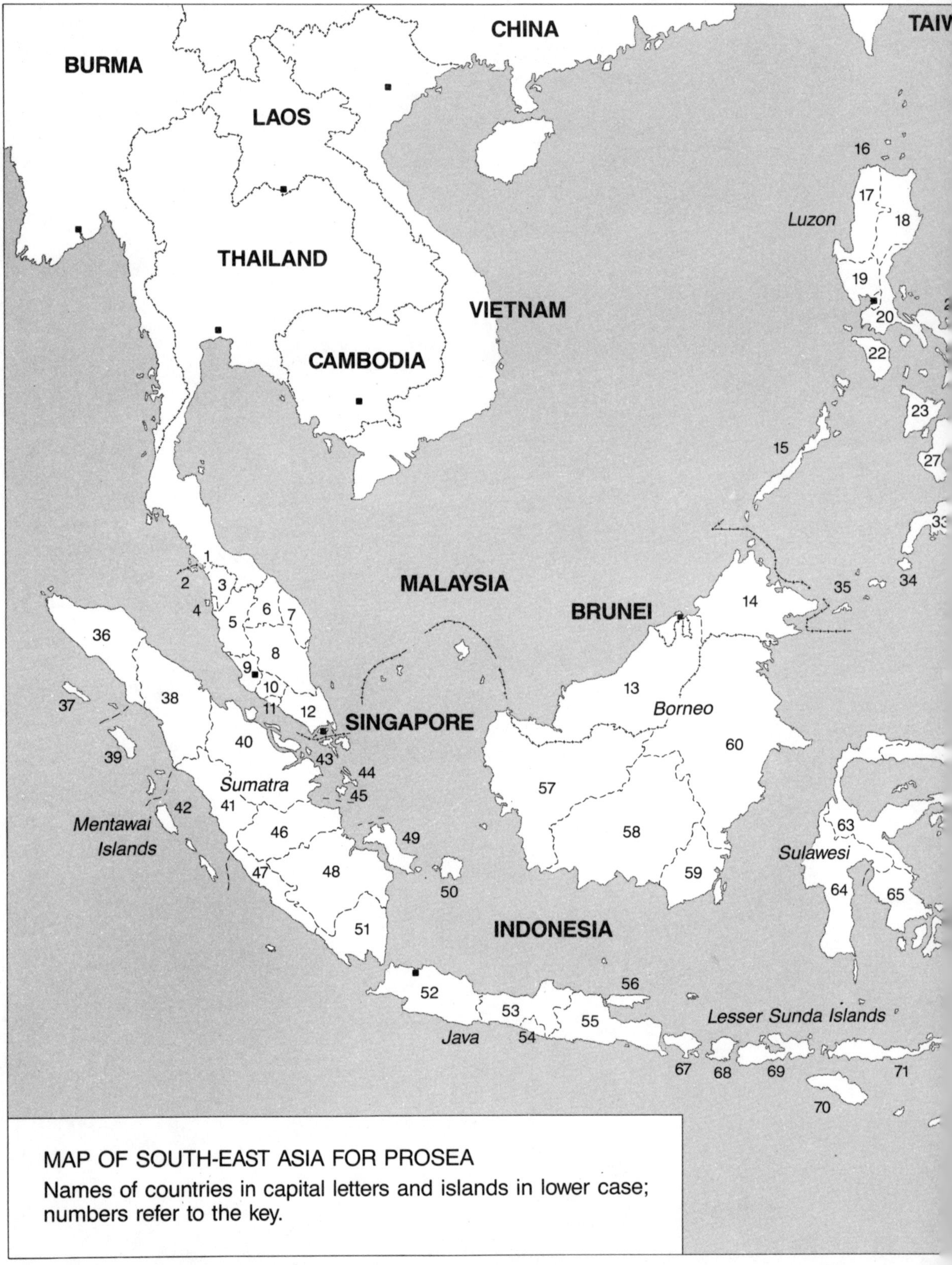

MAP OF SOUTH-EAST ASIA FOR PROSEA

Names of countries in capital letters and islands in lower case; numbers refer to the key.

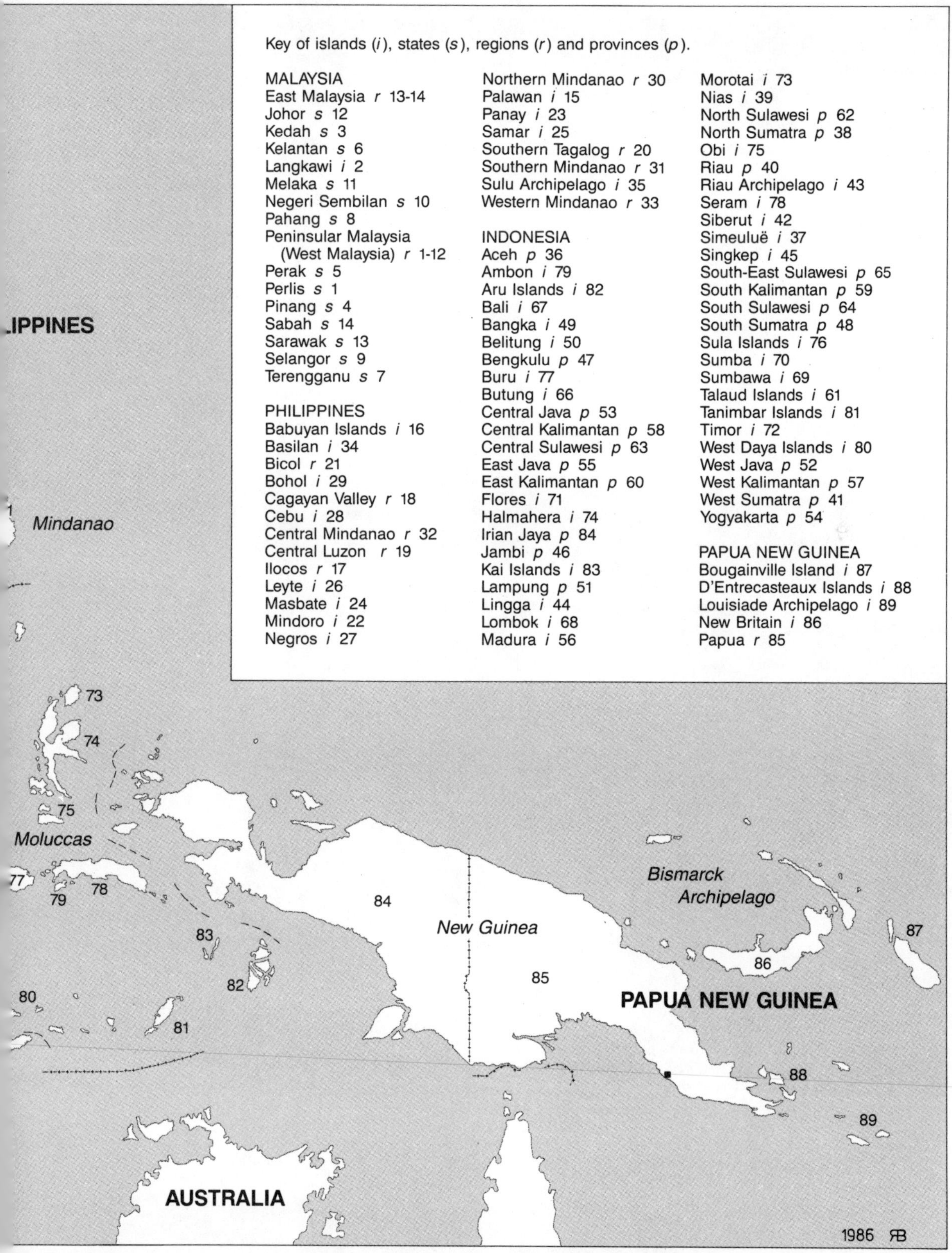

Key of islands (*i*), states (*s*), regions (*r*) and provinces (*p*).

MALAYSIA
East Malaysia *r* 13-14
Johor *s* 12
Kedah *s* 3
Kelantan *s* 6
Langkawi *i* 2
Melaka *s* 11
Negeri Sembilan *s* 10
Pahang *s* 8
Peninsular Malaysia (West Malaysia) *r* 1-12
Perak *s* 5
Perlis *s* 1
Pinang *s* 4
Sabah *s* 14
Sarawak *s* 13
Selangor *s* 9
Terengganu *s* 7

PHILIPPINES
Babuyan Islands *i* 16
Basilan *i* 34
Bicol *r* 21
Bohol *i* 29
Cagayan Valley *r* 18
Cebu *i* 28
Central Mindanao *r* 32
Central Luzon *r* 19
Ilocos *r* 17
Leyte *i* 26
Masbate *i* 24
Mindoro *i* 22
Negros *i* 27
Northern Mindanao *r* 30
Palawan *i* 15
Panay *i* 23
Samar *i* 25
Southern Tagalog *r* 20
Southern Mindanao *r* 31
Sulu Archipelago *i* 35
Western Mindanao *r* 33

INDONESIA
Aceh *p* 36
Ambon *i* 79
Aru Islands *i* 82
Bali *i* 67
Bangka *i* 49
Belitung *i* 50
Bengkulu *p* 47
Buru *i* 77
Butung *i* 66
Central Java *p* 53
Central Kalimantan *p* 58
Central Sulawesi *p* 63
East Java *p* 55
East Kalimantan *p* 60
Flores *i* 71
Halmahera *i* 74
Irian Jaya *p* 84
Jambi *p* 46
Kai Islands *i* 83
Lampung *p* 51
Lingga *i* 44
Lombok *i* 68
Madura *i* 56
Morotai *i* 73
Nias *i* 39
North Sulawesi *p* 62
North Sumatra *p* 38
Obi *i* 75
Riau *p* 40
Riau Archipelago *i* 43
Seram *i* 78
Siberut *i* 42
Simeuluë *i* 37
Singkep *i* 45
South-East Sulawesi *p* 65
South Kalimantan *p* 59
South Sulawesi *p* 64
South Sumatra *p* 48
Sula Islands *i* 76
Sumba *i* 70
Sumbawa *i* 69
Talaud Islands *i* 61
Tanimbar Islands *i* 81
Timor *i* 72
West Daya Islands *i* 80
West Java *p* 52
West Kalimantan *p* 57
West Sumatra *p* 41
Yogyakarta *p* 54

PAPUA NEW GUINEA
Bougainville Island *i* 87
D'Entrecasteaux Islands *i* 88
Louisiade Archipelago *i* 89
New Britain *i* 86
Papua *r* 85